# KNOWING NUKES

WILLIAM
CHALOUPKA

# KNOWING NUKES

*The Politics and Culture of the*

*A          t          o          m*

UNIVERSITY OF MINNESOTA PRESS
*Minneapolis*

Copyright 1992 by the Regents of the University of Minnesota

Published by the University of Minnesota Press
2037 University Avenue Southeast, Minneapolis, MN 55414
Printed in the United States of America on acid-free paper

**Library of Congress Cataloging-in-Publication Data**

Chaloupka, William, 1948-
  Knowing nukes : the politics and culture of the atom / William Chaloupka.
    p.  cm.
  Includes bibliographical references and index.
  ISBN 0-8166-2074-1 (acid-free paper).—ISBN 0-8166-2076-8 (acid-free paper)
  1. Postmodernism.   2. Criticism—Political aspects.   3. Nuclear warfare—Social aspects. I. Title.
  PN98.P67C48   1992
  801'.95—dc20                                91-37209
                                               CIP

To my mother, and in memory of my father

# Contents

Preface      ix

Introduction: Nukes "Я" Us      xi

1. **Knowing Nukes**      1
   Survival—Negation and Unspeakability—A Destabilizing
   Standoff—Discipline and Micro Control—Language/
   Discipline/Politics

2. **No More Warriors**      23
   The Absent Warrior—The Contrary Mission—Demise of
   the Real—Nuclearism/Humanism/Foucault—Replacement
   and Therapy

3. **Robotics (The Bomb's Body)**      43
   The Absent Army—Techno-populism—The "Black Box"
   That Walks—Robot Power—Desire, Absence, and More
   Control—The Darkest Code

4. **Star Wars and the Freeze**      68
   Reaganism—Frozen—Thawing and Other Instabilities—
   C3P0, R2D2, and Darth—All That Is Solid . . .

5. **Immodest Modesty**      86
   Lifestyle Strategies—The Social—Language and Political
   Strategy—Irony and Politics

6. **Power/Cheekiness**      105
   Humanism—Rationalism and "Politics with Attitude"—
   The Ironic, the Abyss—Politics and Open Secrets—
   Sloterdijk's Cynicism—1989

7. **On Lastness: Nuclearism and Modernity**      126

| | |
|---|---|
| Notes | 139 |
| Selected Bibliography | 155 |
| Index | 159 |

# Preface

This book originated in 1987 from my growing feeling that postmodernism—a major theoretical (or antitheoretical) accomplishment of our time—was forfeiting its political role. Since Michel Foucault's death in 1984, few had tried seriously to combine the elements of poststructuralist or deconstructive analysis into a political statement that would continue his serious, unflagging, political commitments.

I never believed that postmodernism was actually apolitical; that common claim never seemed to address fully the serious interventions Foucault and others had generated. In practice—in my classes and in a wide variety of professional encounters—this analytical position had a forceful and thoroughly politicizing effect. The project I began was intended as a political intervention, as confirmation of the extraordinary ability this approach promised—the potential to problematize institutions and practices that had become so resistant to criticism.

Through several interruptions, I kept returning to this project. The end of the Reagan administration and the onset of the Bush administration only confirmed my first judgment. There must be better ways to interact with power in a postmodern era than those we had developed. At the same time, more postmodernists began to write in political terms; both Jean Baudrillard and Peter Sloterdijk, for example, added important dimensions to an emerging political sensibility.

Events, of course, added the most important new elements. The transformations (and near transformations) of 1989 and 1990 pushed me to keep up and to bring these events into my discussion. As I complete most of the manuscript in September, 1990, it is clear that events will proceed; I cannot recap, I can only bring this project to a close and move to another. But I do so knowing that postmodernism will now be a part of any ongoing political analysis of our age.

This project has had more than its share of help. The original essay and the drive toward a larger project emerged from my participation in a National Endowment for the Humanities Summer Seminar under the direction of Murray Edelman at the University of Wisconsin. The faculty and students of the University of Montana provided many opportunities for me to present my work and receive comments. Finally, an earlier version of Chapter 5 appeared in *International Studies Quarterly* 34 (September 1990).

Among the people who made especially significant contributions are Eloise Buker, David Campbell, Gregg Cawley, Scott Daniels, Thomas Dumm, Henry Kariel, Helen Liggett, Timothy Luke, Ronald Perrin, Fran Reader, Mark Reader, Diane Rubenstein, Howard Schwartz, Michael Shapiro, William Stearns, Michel Valentin, and Gary Williams. Several anonymous reviewers added considerably to the project. Errors that remain and omissions that persist are, of course, my own.

# Introduction
# Nukes "Я" Us

*It is of the essence of zeal to be negligent, to believe that what is concealed lies elsewhere, that the past will repeat itself, that the law applies to it, that it is awaited, watched over, spied upon.*

Michel Foucault, *Foucault/Blanchot*

The path to understanding nuclearism can be traced, these days, to unexpected places. Early in 1988, Richland (Washington) High School emerged as just such a place, earning for its efforts a few moments of national attention.

The Richland High community strained to examine key themes of contemporary life, although few there seem to have thought about life in the terms I will use. Ostensibly, the school debated over whether the mushroom cloud was an appropriate "mascot" for the school's sports teams, the Bombers. In doing so, students and teachers wandered into a space where symbols and what they symbolize collide or disconnect, creating gaps in the continuous "fabric of social life" most of us presume exists.

Situated near the Hanford Nuclear Reservation in eastern Washington State, Richland is home to many who have worked on nuclear projects dating back to World War II. Just as a timber-producing town calls its teams the Loggers, or a steel town has its Steelers, Richland (where enriched uranium is crucial to the economy) has its Bombers. In the early 1960s, the school added a visual symbol—a cartoon mushroom cloud emerging through a large "R." *Our* nuke, sky *high*. Nukes "Я" Us.

Vacillating between a naive deconstruction and a sincere pride of place, students at Richland recently debated the mascot. A minority of faculty members—including a baseball coach who had tried to explain the graphic to visiting Japanese students—pushed for its removal. The student body reacted with an overwhelming vote of confidence in the Bombers and the mushroom cloud. As is often the case, however, the process surrounding that vote amounted to a struggle to recapture the power of definition, thus making the package—vote plus explanation—at least partially self-canceling. The student vice president told the *New York Times*, "You've heard of a family crest; our crest is a nuclear bomb."[1] The same student later suggested, "I don't think the student body sees it as a symbol of destruction or weaponry. I don't think it signifies anything other than Richland High School."[2]

In the usual domain of antinuclear thought, we probably would see these students as victims of a pseudomilitary discipline that insists on an inarticulate subject. Still, the students exhibited a certain flair for the paradoxical in their practices of redefinition. Their approach could be reinterpreted (misread, gently) as evoking a deconstructive verve. Surely, it is not an excessive leap to claim that the students were working to come to terms (albeit awkward, even absurdist terms) with an age dominated by strange new twins: nukes and the mass media.

One Richland teacher pronounced that their symbol "stands for nuclear holocaust and the end of the world. It's an inappropriate symbol for high school kids to be glorifying."[3] The students responded by asserting that they were not "a bunch of nukeheads," and giddily embraced a new world. In that emerging world, the ability to work at the level of clashing, contradictory, or even empty symbols is a *positive* characteristic of the exercise of power.

The teachers of these young Bombers taught that the voice of power should be coherent, consistent, and simply persuasive; somehow, their Reagan-era students came to suspect this was not so obviously the case. The students had also begun to suspect that the tools of such power would be taught inadvertently, indirectly. This crucial ability to live amidst the paradoxes and absurdities of a space where symbols and power interact (in multiple, diverse ways) is not something one would approach directly—head-on, looking it "in the eye."

The premise of my book is that we can (and "must," as is often said) find better responses to the nuclear age. The Richland High students may have located some sense of the new possibilities. Faced with a responsibility for the fate of the earth that they did not choose—and would not have chosen—the students acted to exacerbate the gap between symbols and power, not to apologize for it. Having heard, repeatedly, that this fate depended on their solemn renunciations—even though the nuclear age is remarkably immune to such denouncing—the student body turned giddy.

I can identify with the teachers, who must surely be baffled by a generation of students who seem insensitive and disconnected. Still, given the available alternatives, I might join the students. Their refusal to accept symbols as having an obvious meaning (when those symbols are endlessly questionable) situates the students at a key spot, at the end of modernity.

If the choice is to defend absurd responsibility, I might rather be a Richland Bomber.

In the chapters that follow, I intend to deal with the issues raised (even if inadvertently and indirectly) at Richland. In an important sense, nuclearists and their opponents share more than they dispute. Their confidence in a world that passes naturally into speech and writing is one such shared position. The identification of a "values" realm—limited but available for political debate—is another shared position. Each of these agreements (and many others) leaves the institutions, habits, and contexts of authority intact. The Richland students suggest another tack. They direct our attention to the realm of images and problematizations. Following their lead—even if it is naive—we might yet find another way to proceed.

My argument engages the odd conundrums and incongruities of the nuclear age, exhibited on every side of each attempt to politicize or manage nuclear technology and contemporary war. The dominant nuclearism of managers, diplomats, and heads of state has perhaps been too easy a target, in their terminology, and I try to suggest some better ways to read their positions. At the same time, however, the prevailing antinuclear political stance has often started from bad (even if well-intentioned) assumptions, or has misunderstood the efficacy of its best work. What follows, then, is an attempt to begin reworking the discussion that surrounds (and creates) nuclearism.

My approach has an identifiable philosophical background as well, though it may be less familiar to some readers. The linked terms "postmodernism," "poststructuralism," and "deconstruction" have become popular in the North American and European academy, but only recently have they begun to appear in broader public discussions. Without a doubt, they are also overworked, used to simplify arguments that cannot withstand such abridgment. My own project is to begin trying to apply that body of work, applying its general insights about language and society to specific questions in the social world. My presumption is that the oft-repeated criticism of these approaches—namely, that they are inherently apolitical—can best be tested and countered in the context of specific political issues and analyses. Thus, I hope this book will interest those who examine their political lives in view of "postmodernism," but I also intend the book as a setting for theoretical comments about this supposedly apolitical characteristic. More than anything else, the book is an experiment (if only a small, uncontrolled one) in how social criticism and even political science could proceed in modernity's wake.

*Life during nuclearism*—this is a topic that seems to negate the claims of all theorists, especially the sometimes arcane and (supposedly) apolitical crew of literary critics, language theorists, and semioticians. We would, in this view, owe our survival to the pragmatic practitioners of realpolitik, the guys who can strike the deals that keep the missiles from flying. Nonetheless, if my decentered and inconclusive essays in this book have "a point," it is this: while we have been paying serious attention to topics such as "survival," weapons technology, and the (physical) fate of people and earth — apparently on a quest for a metaphysical principle to reestablish a lost center to human life — the era has actually introduced new and unexpected concerns. The seeming "side consequence" of the nuke is that word and meaning are separated even further from each other. The referent (the "reality," the "thing itself") slips away from that which refers to it (the word, the image). Rather than find new faiths that would impose new planetary constraints and codes, we might investigate this slippage of meaning and "fact," perhaps even to discover that this is the most pervasive political development we face. While we have sought to find a new "first principle" — survival, perhaps — another move may already have consolidated its transformation of authority and knowledge.

The application of this theoretical approach to the question of nuclearism has precedents. This book includes a survey of "nuclear criticism," a small body of work emerging from the active scholarly area now included in "literary criticism." This position proceeds from the understanding that (as Jacques Derrida notes) nuclear war is "fabulously textual." One implication of that characteristic "textuality" is that nuclearism is no longer the interpretive province solely of realpolitik. The approaches developed by literary criticism apply, even in world politics — an arena improbably designated the most "real" of policy arenas ever since Machiavelli explained that it was nothing of the sort. In generating that application, nuclear criticism seeks to expose nuclearism's presumptions, but to do so in an unprecedented way, namely, without substituting a new, replacement metaphysics or universalism.

As will become evident, any such project rests on the notion that no code necessarily underlies and unifies seemingly diverse phenomena. This is why it is called *post*modernism or *post*structuralism; structuralisms imply an underlying key to a decodable reality. Marx, Freud, the public-philosophy liberals, and modern social scientists all imply the existence of such a confident project, although some of their confidences are more interesting than others. The response to liberalism or structuralism attempts to reposition reality, values, power, and motivation as diverse, decentered events. As comforting (or convenient) as liberalism, humanism, and a commitment to survival might be, nuclear criticism reminds us that each of these positions also resolves questions before their time, helping us avoid the raging, thorough problematization that resides in any *politicized* realm. Even the humanistic commitment settles important questions too soon, avoiding a challenge to components of the basic, "human" package at its core.

Any commentary that tries to reopen this challenge must, by definition, be continually partial, rhetorical; it will evoke failures of language
and meaning without promising to decode those mysteries. Such analysis,
thus, adopts certain stylistic ploys, as reminders. Seemingly clear conclusions are undercut; puzzles remain unsolved; even authorship becomes an
uncertain claim. The approach is characteristically pervaded by a reluctance to arrive at a final (characteristic) word, a certainty of analysis.

The dips and turns of such an approach are here for a reason; the indirection, playfulness, and sometimes confounding strategies appropriated by any "textual" approach are invoked in an attempt to make
universalisms—now including a universalistic humanism and call for
survival—more unclear, more problematic. Rather than rejecting the universalisms that it criticizes with "new" ones, this kind of analysis probes
political alternatives that continue problematizing "all the way down."
What follows, therefore, is a map that attempts to defy our expectations
of maps. High points, features, and rifts appear, it is true. But what is
most certain about this map is that it shifts unnaturally, reshaping itself
and threatening to dissolve in our hands. The map is partial, incomplete,
provocative; it fails to represent the entire territory. It is a trickster's map.

But if I have tried to produce such a strange map, I must still admit
that it has outlines and borders. In chapter 1, I report on nuclear criticism's intervention into the liberal or humanist approach that thus far
has served as the central code for opponents of nuclearism. The body of
the book is chapters 2 through 5, in which I evoke what social scientists
call (usually with derision) case studies. I would frame them, instead, as
deliberately chosen interventions into crucial paradoxes of nuclearism
and the ways we have tried to live with it. In chapter 2 I consider the residue image of the warrior, the ghost we must impose on nuclearism if the
traditions and continuities of militarism are to remain meaningful. In
chapters 3 and 4 I stage a confrontation between humanist expectations
and technology through an examination of the robot, Star Wars, and the
nuclear freeze. In the final three chapters I consider political possibilities,
in an attempt to frame some elements of a political position that would
privilege the problematizations posed earlier in the book. In chapter 5 I
reread the "lifestyle strategies" activists have deployed, underscored by
the odd slogan, "Think global, act local." In chapter 6 I propose an alternative political attitude, in the wake of this analysis. In the "Last Chapter" (a chapter about "lastness"), I conclude the book, if (I hope) in an
inconclusive way.

A short note on terms is in order. I will use "nuclearism" and its variants
to describe the position taken by the managers and leaders of nuclear
states, even if they seldom identify this as an identifiable political stance.

# 1
# Knowing Nukes

Like few other issues, nuclearism strains to become more than an instance. It aspires to be context and case, to shape public and private life. It seeks a symbolic position of such force that other concerns would arise within the context of nuclear technology, sometimes even when explicit connections are absent. The policies, practices, and discourses of nuclear technology seem to have a capacity to capture attention that rivals even their destructive capability. In short, nuclearism organizes public life and thought so thoroughly that, in another era of political theory, we would analyze it as an ideology. The framework of "survival" or "defense" has become pervasive in Western political cultures, dominating not only the budgets and debates of public life but the more private dimensions as well. In our time, when one dreams of public life, the fantasies may even be atomic.

The level of compulsion attendant to nuclear questions could become a subject of interpretation; a critic could choose to discuss these questions as more fundamental than issues that merely confirm existing frameworks and habits. For citizens of nuclear states, nukes are the metaphors for success and failure, the constraints for experimentation, the analogy for all other "problems." Nonetheless, these same citizens seem reluctant to take nukes so seriously. The background for my project is a suspicion that a sort of conservatism, a slowness to move, characterizes even the most alarmist talk of nukes. The various positions on nuclearism are phrased within familiar political ways of speaking, despite their proponents' considered judgment that precisely these understandings have made the world so different, so dangerous. The nuclearism adopted by states and diplomats presumes a Machiavellian counterbalance of threats, while opponents presume the efficacy of humanist commitment. Despite obvious differences, both positions reinforce contemporary, ideological ways of understanding politics.

Still, there may be openings, hints of recognition that nuclearism is a pattern of its own, not subsumed by existing politics. As the story about Richland High in my introduction might suggest, there are new ways available to think about nuclearism, approaches not yet exhausted by the enormity of the topic. Specifically, there is a lively intellectual approach that has taken a special interest in the question of nuclearism. *Diacritics*, a major journal of literary criticism, published the symposium that gave this approach the name "nuclear criticism." Major figures such as Jacques Derrida, Michel Foucault, and Jean Baudrillard have used nukes explicitly as instances of their work, and many other authors have addressed nuclearism in the context of discussions about texts, discourses, and authority.

But, as predictable as a contemporary intellectual's interest in nuclearism might be, there are resistances to remember as well. Nuclearism turns back on such intellectual efforts, carrying special implications and challenges for them, as do only a few other political concerns.[1] Given the generally perceived seriousness of nukes, the playfulness and studied irreverence of the language approach seem blatantly inappropriate. Nuclearism obviously confronts deconstruction's critique of totalities, of global ways of understanding politics. Political talk of nukes, on all sides, constantly moves to universalistic levels, almost as if there were a collective wish to replace totalities that no longer hold the spell they once did. Nuclear criticism clearly (or, as Derrida notes, nu-clearly) raises a range of other possibilities. In short, I am trying to portray nuclearism and textual/interpretive approaches as key examples of each other, as an intersection where difference produces previously unclaimed space for political action.

In this chapter, I begin to map that intersection by examining a key universalism contained in most of the politics that engage issues of nuclearism. Survival is a coded position that privileges certain questions and marginalizes others. In this chapter, I want to make aspects of that privilege more explicit, more accessible to controversy. Assigning "survival" the status of summary and goal implies a relatively settled "humanity" that is, one hopes, to survive. One would hardly need to hope that people not survive to bring that code into question (although some radically misanthropic environmentalists playfully take just such a position). Savvy about codes and symbols, the nuclear critic might begin by highlighting what our most universalistic survival codes assume.

## Survival

Perhaps the central political metaphor of antinuclearists involves the appeal to survival of the human species as a principle that can guide social and political response. But after Jonathan Schell's *Fate of the Earth* presented that case, the narrowness of this appeal began to draw criticism.[2] As Robert Jay Lifton has noted, the emergence of neo-Nazi survivalists is not without consequence for nuclear opponents who have used the survival language so extensively themselves.[3] Schell broadened his metaphors in *The Abolition*,[4] and nuclear opponents in general have tried to define survival in a way that is not individualist.

Nonetheless, recent essays by political theorist George Kateb bring even that modified project into question, finding within the "survival" position an indefensible replacement totality.[5] Kateb's critique focuses on the political metaphysics implied by the survival position. To turn "existence" into a principle that could inform action is to ignore many other philosophical commitments made in this century. The metaphysical privileging of existence as key to a great and total meaning (that might motivate political action in a classically liberal framework) is unavailable "in an age when the death of God has been announced with adequate plausibility."[6]

> Existence does not have systemic attributes amenable to univocal judgments. At least some of us cannot accept the validity of revelation, or play on ourselves the Kantian trick of regarding existence *as if* it were the designed work of a personal God, or presume to call it good, and bless it as if it were the existence we would have created if we had the power, and think that it therefore deserves to exist and is justifiable just as it is. No: these argumentative moves are bad moves; they are transparent tricks.[7]

Kateb wants to articulate a defensible "attachment to existence" without relying on "any kind of totality." Existence

> cannot be justified by any "internal" or human standard developed independently of a supposed divine authentication. That is to say, attachment cannot be cultivated by way of a theology . . . or by way of a believable reconciliation to the facts of wickedness, suffering, waste, cruelty, obscenity, and death. The universe . . . is without sponsorship; and existence on earth fails every test that is strenuously pressed. . . . What is needed is precisely a mode that is content not to make the world—human and natural existence on earth—into a story, a picture, an order or a pattern . . . that is, into a self-adequate totality or into a necessary part of a transcendent totality.[8]

The puzzle we retain, after Nietzsche, is to find a way to establish human value without the aid of an external totality (whether religious, scientific, or merely commonsensical).[9] Kateb's strategy is to shift attention to the institutional and philosophical contexts within which this discussion of survival proceeds. Such a broadening of the question could confront the excessive individualism that otherwise makes "survival" a suspect theme. If the extreme individualism of this century cannot be absorbed into metaphysics, Kateb claims, it is still the case that individualism and the institutions of democracy are not easily dismissed: "Individualism in some of its developments after the seventeenth century contains . . . saving thoughts and feelings. The great work of Emerson, Thoreau, and Whitman comprises the main development, and the phrase 'democratic individuality' perhaps best names their idealism."[10]

To broaden these saving possibilities within individualism, Kateb (in a surprising move, for him) suggests that we take heart from the "antidemocratic individualist doctrines of Nietzsche and Heidegger," both writers who influence contemporary language approaches. "The best defensive idealism is individualism," but "the self-surpassing of both rights-based individualism and existentialist individualism is the unique source of a selfless and saving attachment."[11] In short, without adopting the categories I am attaching to this position, Kateb poses the possibility of a historical, yet ambivalent and even poststructuralist, individualism—a political form of nuclear criticism—as a response to the broad crisis of meaning in the late modern era.[12]

Kateb distinguishes himself from Foucault and Derrida when he stresses continuities, arguing that the dilemma for democratic individualism was highlighted by the nuclear age, but was visible previously. Several features of democratic society have long been at tension with the democratic idealism he sees as that society's best protector. Not only this puzzle, but also its resolution, precede nukes. Citing Whitman, Kateb argues that a conventional, democratic individualism could be founded on practice. Being " 'both in and out of the game and watching and wondering at it' . . . is far better than being rooted in what is superstitiously regarded as reality. To watch the action as one acts is to play; to play is never to lose sight of others."[13]

This amounts to a proposal for an individualism defined relatively, justified by social and aesthetic judgments, and implemented on the model of play. Even after the metaphysics of existence has become impossible, politics and thought continue, because "democratic individuality radically changes both action and contemplation."[14] Whether or not Kateb specifically intends it as such, this is a poststructuralist reading of democratic individuality—pragmatic, aesthetic, and interpretive. The individual acquires the critical distance necessary to judge his or her existence by *acknowledging* that the external vantage point previously provided by theology is now absent. On the basis of that understanding, the episte-mological standpoint of individual thought and action can shift. Sources of meaning dislodge from supposedly essential, natural "facts" of exist-ence, and instead situate themselves in the interplay of contemplation and action. Thus, "the hidden source of modern democracy may always have been the death of God." But the (nuke-induced) "precariousness of exist-ence now deepens this sense," moving us toward a preferable *democratic* possibility. "Individualism in its contradictory variety is the best defen-sive idealism in the nuclear situation."[15] Whether or not this is indeed an "idealism" is an issue nuclear criticism would pose to Kateb. In any case, it is an unlikely "idealism" — lacking ideals or a positively structured given practice in which to situate them.

Kateb's analysis may be most useful for nuclear criticism's deconstruc-tion of "survival" as a cornerstone of this debate. Still, others interested in the relationships between language and politics will be dubious about Kateb's defense of "idealism," with its implication that a strong role still exists for the intellectual as a speaker of that ideal, a judge of its cases, and an articulator of what "existence" and "survival" might *be*. Ironi-cally, Michel Foucault's treatment of this possibility also includes this approach's first insights on nuclearism:

> Some years have now passed since the intellectual was called upon to play this role. A new mode of the 'connection between theory and practice' has been established. Intellectuals have got used to working, not in the modality of the 'universal', the 'exemplary', the 'just-and-true-for-all', but within specific sectors, at the precise points where their own conditions of life or work situate them (housing, the hospital, the asylum, the laboratory, the university, family and sexual relations).[16]

This passage should remind us of the roles played by intellectuals in the nuclear opposition. Humanists have learned the physics of power plants in order to object at siting hearings near their cities. Physicists, simulta-neously, have learned the language of political opposition, organizing col-leagues against Star Wars in their universities and institutes. In the exam-ple I will consider in a later chapter, the intellectual contribution was a phrase (the nuclear freeze) and a strategic political approach — not a man-ifesto of values and ideals.

In the interview quoted above, Foucault goes on to make his best-known comments on nuclear politics. His claim is that the intellectual *par excellence* is no longer the writer, who brings that "idealism" to concrete form, but the university activist, the "technician, magistrate, teacher." Global significance is not lost in this transformation. Such actors "have become able to participate, both within their own fields and through mu-tual exchange and support, in a global process of politicisation of intellectuals."[17] Foucault's example of an intellectual who operates in the realm of the specific is a central nuclearist:

> This figure of the 'specific' intellectual has emerged since the Second World War. Perhaps it was the atomic scientist (in a word, or rather a name: Oppenheimer) who acted as the point of transition between the universal and the specific intellectual. It's because he had a direct and localised relation to scientific knowledge and institutions that the atomic scientist could make his intervention; but, since the nuclear threat affected the whole human race and the fate of the world, his discourse could at the same time be the discourse of the universal.[18]

Focusing on the discontinuity entailed by nuclear technology (rather than on the search for continuities, as Kateb does), Foucault reconciled the role of the intellectual with the epistemological break required for "survival" to make sense as a political position. In a genealogy of nukes, the displacement of survival as key concept may be the crucial move toward oppositional politics. At least, that displacement marks the seriousness of the break with previous stances. Without that break, "survival" represents, at best, an appeal to a philosophically precarious doctrine of existence. At worst, it could be a selfish preference, little more than a narrowly narcissistic concern for physical health.

Survivalists of every political stripe would respond that there is a general issue at stake, whether we like that issue, or whether the philosophical or psychological dimensions of that issue are felicitous or not. In other words, they are appealing to a brute condition, a stark threat that we cannot choose to ignore. The nuke—in league with the antinuke— does make it plain that we have common "species" interests, as the survivalists argue. But the issue is still not that simple. On one hand, this claim of species interest must confront the possibility that it is a vain or opportunistic claim. That is to say, it is not a self-evident condition. The concept of a self-aware species is a political act, inextricably bound to the possibility of political response—the possibility that all survivalist politics requires. On the other hand, such a position also must confront the fact that this species constitutes itself by identifying interests and solutions; there really is no "ordinary life" to return to after we settle survival issues. That political struggle already will have conditioned whatever life one would then resume. In other words, the species may have interests, but it is also the case that such a species is constituted, not found or remembered. In short, the call to survival not only addresses "real" lives (whatever those might be), but also constitutes those lives.

What does it matter that this constituting activity has happened? Crucially, this constituted species sees itself as natural (what else could a species be?), but that perception is at odds with its situation. The context is far from "natural" (in the sense that no strong coherence underlies it); a better case can be made that it is contrived, contradictory, rule-bound, and, finally, absurd. Foucault's accomplishment, then, was not only to have joined with existentialists, Dadaists, and others who have so effectively "denaturalized" human history in this century. In addition, Foucault advanced these efforts by showing possibilities for freeing activity available only after history is denatured. For the species to act on the goal of survival embroils us in a simplistic, if still powerful, circle. The species must have always had some motivation to survive as a species, but its commitment to certain practices (especially rationality and science) is both unquestionable and the source of the threat amidst which the species finds itself lodged. Thus, the species must have mutated to produce such a result, and a mutated species might not be able to act on behalf of its survival. The absolutization of humanity proposes to lead us away from the twists, perversities, and gaps that continually preside over the nuclear age. Absurdity and contradiction have become elemental terms in our era. They are "hardened positions," to borrow a term, even if the notion of a hardened irony might be familiar only to Baudrillard.

The species survival position cannot be comfortable in emphasizing those absurdities. But unless it does so, the survival position can scarcely discuss the nuclear age at all. From the approach I am taking, then, we might even call this diagnosis of unspeakability a rhetorically determined stance; antinuclearists have been forced to describe the age as unspeakable in order to continue to draw upon and defend an absolutized, natural humanity. As a consequence, the species survival position may not notice the broad effects of the age's distinctively *spoken* (speakable) character.

Nuclear criticism could offer a better political response if it could expose the specific operations of power that enable some politics of opposition. Before considering that possibility, however, we must be more precise about this "unspeakability" that continually haunts talk of nukes.

## Negation and Unspeakability

If survival has been the telos of nuclear opposition, its *raison d'être* emerges from the claim that nuclearism represents such total negation that it is simply, even unspeakably, dominant. Just as Marx identified the moment of negation for capitalism (revolution), and Locke identified the same moment for feudal society (individual liberties), writers such as Lifton argue that nuclear culture implies not simply danger but also a profound, cultural negation. As Barry Cooper writes,

> The sheer destructive power of atomic weapons has ensured that not
> even fraudulent recognition could ever be gained through their employ,

and this is known to everybody beforehand. As one *Wibakusha*, or
A-bomb survivor, observed: "Such a weapon has the power to make
everything into nothing." Nuclear weapons do not simply destroy
things, they destroy the boundaries between destruction and non-
destruction. The destruction . . . at Hiroshima, wrote Lifton, was so
nearly total and long-lasting that "the survivors of the bomb have
experienced a permanent encounter with death."[19]

Amidst such an encounter, it has been commonplace to assert that we
are in the realm of the "unspeakable." In short, nuclear opponents im-
plicitly admit that nuclear war is a representation, then put that image in
a rhetorical context that evokes (represents) the most profound absence
possible. The nuke implies a prospect for such thorough annihilation that
it is "unspeakable," an image of future negation so total as to illuminate
a sort of new, negative totality on which to base political action.

While opponents have tried to balance unspeakability with vociferous
protest, they have seldom considered seriously another possibility:
namely, that a lively opposition to nuclearism might arise along a different strategic tack, one that dislodges the whole presumption of unspeakability. Jacques Derrida's deconstruction of "unspeakability" not only
marks the limits of antinuclearism; it pushes opposition to nukes toward
this other possibility, which is completely within his approach.[20] Far
from "unspeakable," Derrida argues, nuclearism is uniquely textual; that
is, it is "spoken" in a sense that other political quandaries have never
been.[21] Nuclear criticism finds a topic "whose essential feature is that of
being *fabulously textual*, through and through."[22] Information, communication, codes and decoding thoroughly pervade nuclear politics. What
is more, this political position is ahistorical in a way that confirms the
deconstructionist claim that history cannot be relied upon as a strongly
meaningful political precedent. As Derrida notes, "A nuclear war has not
taken place: one can only talk and write about it." Other wars were only
talked about and written about before they happened, but none of them
marked this kind of break.[23]

> Unlike the other wars, which have all been preceded by wars of more or
> less the same type in human memory . . . nuclear war has no precedent.
> It has never occurred, itself; it is a non-event. The explosion of
> American bombs in 1945 ended a "classical," conventional war; it did
> not set off a nuclear war. . . . Some might call it a fable, then, a pure
> invention: in the sense in which it is said that a myth, an image, a
> fiction, a utopia, a rhetorical figure, a fantasy, a phantasm, are
> inventions.[24]

Peter Schwenger has added to this notion of the textuality of nukes by
reminding us how literary their invention was.[25] As he notes, H. G. Wells

not only predicted the bomb; he also coined the name "atomic bomb," dedicated his 1913 book, *The World Set Free*, to the physicist Frederick Soddy (whose work would be important to weapons physics), and predicted the year (1933) when the artificial radioactivity that eventually led to the bomb would be produced. Leo Szilard, the physicist who accomplished that "real" chain reaction, had read Wells's book in 1932; Szilard introduced his theories with references to Wells, then dedicated himself to the world-government politics Wells had also predicted.

Both Derrida's insight and Schwenger's anecdote invite the opening of a whole realm of oppositional activity, of which only a few examples now exist. The premise of this genre ("speaking unspeakables"), as Derrida claims, may have been best realized before the nuclear era, in the literary texts of Mallarmé, Kafka, or Joyce. But there have been contemporary attempts that nuclear criticism could address.[26] One could imagine a comparison, for example, of two highly publicized television films of the Reagan era, "The Day After" and the right-wing response to it, "Amerika." The level and ferocity of the response suggest that "The Day After" broke a taboo. "Amerika" charges weakness, appeasement, and even collaboration, but these charges so completely miss their target that we search for a better interpretation. Perhaps "The Day After" transgressed in a special way, and the only available way of responding was the arcane code of anticommunism. The actual taboo it broke, it broke by speaking *at all.*

At the same time, the activity of finding new ways to read (literary or cinematic) texts about nukes must relate to the broader project of empowering responses if such activity is to fit within the antinuclear schema I am discussing. Leaping over hypothetical psychological diagnoses to speak *politically*, such a development is not so hard to imagine. "Speaking the unspeakable" has never been a happy entrée into activism. Nuclear opponents have adopted any number of rhetorical strategies for overcoming this obstacle. They argue that this "unspeakability" denotes an importance so huge that we must dissolve the reticence and disgust that is our "first reaction." Or, alternatively, they dissolve their political position into a therapeutic one, implying that the contemporary citizen would be healthier and less conflicted if she would admit and confront the nuclear demon. In either case, the political use of unspeakability produces a paradoxical stance at odds with the naturalism of the survivalist, species-interest position.

This unacknowledged (unacknowledgeable) taste for paradox goes even a step further. Having bound themselves in multiple, endlessly and effortlessly proliferating dilemmas, nuclear opponents then announce that it is their goal to impose the condition of "unspeakability" on nuclear managers. The solution to the paradox of nuclear strategy is to silence strategists, such as Caspar Weinberger, who dare to speak of limited nuclear war. This enforced silence has long since ceased to be uncomfortable for nuclear managers, who now clearly understand that their control will proceed more satisfactorily when it is invisible. Opponents, then, have undertaken the odd project of enforcing unspeakability, on the one hand, while also seeking to make nukes *visible*, thus making them controversial—a topic of conversation.[27]

Such strategies have a validity, as I will discuss in a later chapter, but it is not necessarily the validity the opponents promote. Just making the artifacts of nuclearism visible isn't enough; they don't speak for themselves. These artifacts—whether warheads or power plants—surely offer little help out of the paradox of unspeakability that both veils and unveils them, and all the while also seems to expect a solution. Finding nukes not only "speakable" but "fabulously textual," nuclear criticism can respond to this odd political situation in part because many more strategic approaches become possible once we move the response to paradox out of an "unspeakable discourse" and into a textual or literary context.

Frances Ferguson has attempted just such a move, beyond the practical or ideological. Doing textual criticism of both a standard homeowner's insurance policy and (at the other end of the scale of gravity) Schell's *The Fate of the Earth*, Ferguson argues that neither succeeds in its attempts to "think the unthinkable."[28] Other genres are activated—in Ferguson's case, the genre of the sublime—once we understand that we are in the realm of texts. Nuclear critics have already shown that many other literary devices also pervade this discussion. Their work has already confirmed Derrida's deconstruction, showing that the seemingly simple antinuclearist faith (contra unspeakability) has been substituted for political judgment of events, of the movements of power.

There is, to be sure, another way to take Derrida's intervention into nuclear discourse, a way that remains grounded in some version of "reality" while still preserving space for interpretation (and hence "speakability"). This is the "reconstructive" project, and it is sure to become fashionable, in the face of deconstructive approaches that resist criticism as well as they do. Surely there is a way to "split the difference," to get on with the intellectual and liberal tradition of applying values to political issues in the name of ethics. Perhaps the best example of such an attempt is found in J. Fisher Solomon's *Discourse and Reference in the Nuclear Age*:

> What I am suggesting . . . is that our interpretive activities are not
> *wholly* rhetorical. Rather, they are conditioned by something that
> exceeds discourse, that exceeds tropology and difference, and that, while

not reducible to an absolutely determinate identity (dialectical or otherwise), is not to be found in the figure of differance.[29]

Hence, Solomon can join the ranks of those who have tried to revive the Peloponnesian War to understand international politics, asking, "Do not such political circumstances as those that led to the outbreak of the Peloponnesian War bear a certain empirical potentiality, a certain propensity for conflict, that we can take into account as we seek to interpret the course of similar events?" This leads Solomon to Aristotle in the attempt to locate an "ontology of potentiality and actuality," a "potentialist realism."[30]

Solomon's treatment of Aristotle is careful and serious, avoiding the "organicist and transcendental theology that is virtually [?] impossible to accept or to defend," while still pursuing an "ontic potentiality."[31] When Solomon finally gets to Foucault—who might have helped to clarify the nexus of potential, specificity, and language—he cites only one early collection, and immediately turns back to a discussion of what Derrida might or might not permit in the political world.[32] At some points, Solomon resorts to familiar criticisms of poststructuralism, letting show his reluctance to take Derrida all *that* seriously. At one point, literary discourse is contrasted to politics on the basis that political action seeks to interpret in order to discover "the concealed meanings, the real references, behind the rhetoric."[33] It is by now well established that literary intervention can do more than this, while still remaining fully political. Or, to put it another way, to limit politics to the interpretation of concealed meanings is to prevent prospective political actors from engaging in the world as it has emerged.

Solomon moves (in a perfect postmodern trajectory he also denies) from Aristotle directly to Heisenberg and Popper, then to $L = A = N = G = U = A = G = E$ poet Ron Silliman, and on to reader response theory and Saussure. But even as he makes all those agile turns, the distinction between potential and metaphysical has some difficulty catching up to the thoroughly discursive context Derrida and others have already established. Still, even by Solomon's standard, the intellectual movements surrounding literary criticism have begun a major intervention against the supposed "unspeakability" of nuclearism.

When nuclear critics begin breaking the hold of this (inverted) injunction not to speak, we may begin to suspect that at least some antinuclear approaches may have served the interests of contemporary, disciplinary power. When antinuclear activists issued their calls to action, something was always amiss; their claims announced importance when no such announcement was needed, given our obsession with nukes. The meaning of this "surplus denotation" should always have been obvious; "not speaking" has conveniently covered the movement of control. It has directly served nuclear managers whose discourse is so precarious, despite its importance and their evident confidence in it, that it cannot tolerate discussion. In other words, the reconstruction has been initiated too soon, before the terms of debate were quite understood. Survival worked poorly as a basis for reconstruction; something else—even closer to meaning, expression, and politics—had been at stake all along.

Having deeded control of the nuke to managers and technocrats, who assured us that their approach was structurally capable of dealing with complexity, speed, and radical uncertainty, we discover—at the end of their line—that it was always their control that was at stake. Politics reentered the scene at the unlikely hands of transitional figures like Reagan and Gorbachev, and the negotiability of the nuclear age almost instantly became obvious to everyone.

## A Destabilizing Standoff

Although opposition has focused on the "madness of MAD," literary criticism has available the analytical tools to demonstrate instability more convincingly. The paradoxical military mission (to prepare weapons so that they will never be used) may have been more difficult for civilian and military leaders to maintain than even their opponents realize. Deterrence itself implies a conjuring of power—a fable of power. Never used but always effective, the power of the nuclearists could be seen as the greatest single accomplishment of the poststructuralist era. Michael McCanles traces this type of power back to Machiavelli who, McCanles argues, may have been the first to discuss "an aspect of human behavior that has become the focus of theoretical attention . . . [only] since World War II: the assimilation of human conflict to hypothetical models of games and strategy."[34] The particular brand of hypothetical modeling favored by the nuclear strategists forces emphasis on an ultimate "destabilizer"—nuclear war—while ignoring or benefiting from any number of other destabilizing events. Perhaps the most relevant of these excluded events is the one introduced by this schema of models and games itself; in play, the players will test limits and explore possibilities. The modeling activity itself is the great destabilizer, even if it proceeds under the sign of stability.

Furthermore, the swirl of interpretations, paradoxes, and fears has a predictable structure. Within this particular game, this is not only the well-known "standoff" structure, either: as McCanles notes, a certain entropy is just as characteristic of this game as any standoff.

> Recognition of the paradox . . . will inevitably generate understanding
> of the entropy of the threat. This understanding will in turn generate the

paradoxical coexistence of equilibrium and destabilization of
equilibrium, and both sides will continue to entangle each other in this
paradox by [accusing] each other of bad faith and [justifying] further
metastrategic plans for nuclear build-up.[35]

Even our best signs of stability are easily inverted into signs of chaos and
entropy. The only dependable stasis refuses to be static; interpretation demands a role, despite our wishes it would recede.

This absurd outcome may be most evident when we consider those
major destabilizers in the nuclear world that come under the classifica-
tion of "accidents." The term "accident" is of obvious interest to nuclear
criticism. In a discourse that allocates responsibilities pervasively, "accident" is a free spot, without cause or conspiracy. In the case of nuclear
power, the notion of accident had already become visible in the late
1970s, after nuclear critics and Nuclear Regulatory Commission officials
sparred over the vocabulary appropriate to Three Mile Island. To official-
dom, accident was obviously an appropriate label for these events, since
there was never any suggestion of malevolence or subversion. To critics, it
was just as obvious that when societies produce electricity by placing or-
nately complex plants around the landscape, radiation releases are so inevitable that the word "accident" reveals an evasion of responsibility.

In another case, compatriots of the Iran Air 655 victims insisted that
its destruction must have been intentional, simply because the powerful
American technology could not possibly have "made a mistake" (or "had
an accident") of such magnitude. Meanwhile, critics in the United
States—more familiar with technological failures—argued that placing a
weapon such as the U.S.S. Vincennes in a place such as the Persian Gulf
invited tragedy so openly as to defy the categories "mistake" and "accident." Noting the radical reversibility of such analyses—the ease with
which they are inverted—we might begin to suspect that "accident" is a
special term in the debate over nukes. Indeed, "accident" has even served
as a sign of stability, as in the oft-repeated analysis that the paradoxes of
deterrence are so stable that the real danger of nuclear war comes from
the chance of accident.

So-called accidents may attain this special status because of the role
the rhetoric of "accident" necessarily preserves for a rhetoric of agency.
To call something an "accident" is to claim (or hope) that there is no har-
bor for responsibility, even though we continually use rhetorical devices
that allocate causality when we talk about politics. This double character
gives the formulation "nuclear accident" an extraordinary power. Hypo-
thetically, such an accident could destroy all life; if that weren't enough,
the formulation draws attention to the provisional, constituted character
of American discourse about agency and authority.

Richard Klein and William B. Warner presented the Korean Air Lines downing as a case that illustrates the ambivalence of accidents.[36] As they suggest, we have long known that designating something an "accident" is an implement of international diplomacy. Such a designation can be (and often is) constructed after the event in question, for purposes not necessarily connected to the "facts" of the event. Statesmen make events into accidents (or, conversely, attribute a conscious purpose to an inadvertent event) depending on the geopolitical move they want to make. In the case of nuclear war, which has no "after the fact," these determinations would have to be made very quickly, and "this determination of the character of the incident, before it happens, may itself initiate a war."[37] In such a situation, it might well be impossible for the participants to map all of the contingencies required to produce reliable clarity. Indeed, clarity on causation, responsibility, and accident has often been an artifact of "the luxurious time of diplomatic distance," not some obvious feature of the event in question. And clarity, as it pertains to nukes, is no abstract exercise; it is a precondition for continuing at all. One failure and the rubble bounces, as the saying goes.

The case of KAL 007 is illustrative. This time, there was an "after," so we have the usual and predictable diplomatic interpretations to examine. The Soviets cried foul, charging conspiracy. The Reagan administration renewed its claim that the U.S.S.R. was an evil empire and used the event to justify weapons requests. No surprises. Klein and Warner's point, however, is that in the heat of this particular night, it would not be even slightly implausible to suggest that hugely different interpretations of this event could have prevailed in Washington and Moscow, whatever the "actual" facts and motivations were. This is an interpretive moment, and these interpretations tend to diverge, not to converge in some safe and reassuring way.

> From the Soviet vantage point it hardly seems an accident that the course of KAL 007 happened to coincide with the course of a U.S. RC-135 spy plane. But from the vantage point of the U.S., the flight "deviation" of this particular plane does not seem so surprising at all; it may in fact be inevitable given the thousands of flights along this Pacific route. . . . Thus, what seems a telling coincidence to the collective subjectivity defined by Soviet leadership seems merely accidental to observers . . . who do not share the same national subjectivity.[38]

Klein and Warner use literary interpretations to show how utterly incomprehensible this "fact" may have been in its unfolding. One can even imagine that KAL 007's James Bond–like name imparted confusion. That name could have been seen as proof that this was no spy mission (obviously, they wouldn't have named it *that*), or proof that it was such a mission (they'd never suspect *this*), or evidence or a classic spy's slip, betrayed by "what he has taken every conceivable rational precaution to conceal."[39] The indeterminacy of language and the characteristically linguistic, interpretive nature of such politics take away any reassurance we could be offered that, despite all our critical complaints, we have only "accidents" to fear now.

Or, in slightly different form, we can imagine an interpretive moment — fraught with levels and complexities — far more difficult, even, than an episode in which one had to "get the facts." "The injured party will not enjoy the luxurious time of diplomatic distance from the event that allows one to choose" a course of action. Instead, the injured party finds himself in an almost inevitably catastrophic position, trying "to determine in these swiftly passing moments, before the end, whether he is not actually already at war," knowing, perhaps, that his attempts to determine "the character of the incident . . . may itself initiate a war."[40] To demarcate something as "an accident" is to imply that it is outside the rationalist realm of planning and decision that supposedly lies at the core of the national defense. Actual events, however, fail to honor such demarcations; a successful political actor manipulates them and gains benefit.

The "accident," then, exposes the presumptions of nuclearist positions that propose that such events are all that remain to fear. Indeed, we should have long ago seen through the rhetoric of "accident." As Garry Wills has explained, the entire nuclearist project suffers from a reversal of Clausewitz, who "understood that the very conditions of war tend to break down the effective conduct of war."[41] Presuming that "everything works" ignores Clausewitz's advice that a sizable margin of error must be assumed. On the battlefield, even the most dependable moves will break down. "Danger, of itself, takes a toll, in apprehension or despair, in heightened alertness or the racing of one's pulse. And danger, says Clausewitz, is the very air one breathes in war. It charges the atmosphere, giddying a person, unsettling judgment."[42] Nuclear strategy has veered sharply away from the master strategist's insight, even while our intimacy with danger has intensified.[43] Not only do we presume that our devices will work (and SDI raises that presumption to new levels), we even base our strategy — in the case of "window of vulnerability" scenarios — on the assumption that the Soviets also will act on the assumption that their own weaponry is infallible.[44] Seeking managerial control in the form of deterrence, nuclearism strays off course, elevating the "accident" to a new, reified status. In this new context, accidents will happen — continually taunting the managers' forgetfulness of Clausewitz's most obvious points.

It is not technological bugs, then, that deliver us to perilous times, so much as it is confusions of agency and misunderstandings about the role of plans and strategies. Citizens and nuclear strategists alike have blithely ignored some long-understood tenets of politics and war, and the traces of that forgetfulness can be identified within nuclearist discourse itself, as the case of "accident" shows. This gives the era a hallucinatory quality, when the master-in-control reveals his own foibles. And, as Klein and Warner conclude, "Hallucinatory effects and effects of coincidence acquire, in this space, uncanny power to become the bases for fateful decisions."[45]

## Discipline and Micro Control

It may be surprising, given the liberal basis of much nuclear opposition, that the issue of control is not one on which pro- and antinuke positions are always clearly distinguishable. Some nuclear opponents have used the threat of annihilation to *justify* control measures, arguing that the controls of the nuclear managers are inadequate. More often, of course, critics cite this control as a reason to eliminate nukes. Critics argue that the amount of citizen control necessary to secure the safety of nuclear civilization is so odious that nukes—rather than safeguards—should be foregone. Once again, we find ourselves in the middle of a strange political situation. Opponents argue for controls (intervening in plant siting proceedings and urging that similar procedures apply to weapon installations), but also use these controls as an argument against nukes. Nuclear managers justify widespread surveillance and disciplinary measures by their observation that the world is so very dangerous, but they then argue that deterrence has stabilized the world under their leadership, and that this action somehow relates to "freedom." Repeatedly, the sign of the paradox presents itself as the characteristic sign of an era that strains to ignore those signs and to present a politics of values in response.

Trying to master such awesome exercises as "thinking the unthinkable," we are susceptible to seemingly small, even trivial, moves of power. Discipline, already at a huge advantage, conceals itself further. The shift of politics into techniques of control accumulates beneath this weighty umbrella of "big" concerns. Political micro controls parallel the claims of nuclear managers, adding stability to every power claim made on behalf of the state. On one level, disciplinary measures are the predictable accompaniment of having so many dangerous weapons and associated facilities. At this straightforward level, it would not be difficult to foresee the need for control. The most famous case of such foresight is probably Einstein's. Within a year after Hiroshima, he imagined a situation that "[would] insidiously change men's lives more completely than did Hitler"; "To retain even a temporary security in an age of total war, government will have to secure total control. Restrictive measures will be required by the necessities of the situation, not through the conspiracy of willful men.[46]

Einstein was correct on oppression and conspiracy, but he was wrong on power. Nuclearism has found a way to spread that power throughout society *without acting*. This is Jean Baudrillard's notion of contemporary power. "The nuclear threat is part of a 'soft' mode of extermination, bit by bit, by deterrence, not at all an apocalyptic term."[47] Such power implodes and leukemizes, in Baudrillard's metaphors; its focus is inward. In this instance of power (which clarifies and exemplifies Foucault's controversial argument), there is no action, intention, or conspiracy. Never using its weapons—indeed, founded on the notion that the best force is never used—power can infect everything. It is discursive in the sense suggested by Derrida; this power is only spoken, never used. Outside history, beyond the world of things that have functions and uses, this is precisely poststructuralist power.

Baudrillard's alerts and Derrida's observation on textuality share a poststructuralist position on power. Earlier, Foucault had identified sexuality as an area of activity that was supposedly repressed (perhaps even "unspoken") but was, in fact, pervasively, compulsively spoken. Indeed, that supposedly "natural" realm of activity has been, he claimed, radically constituted in discourse. There are parallels with nukes. We speak of a repressive relationship with nukes; we are said to have repressed our anger at the nuke, or our awareness of it. Lifton gave this psychological phenomenon the name "numbing," which is one of the central theses of contemporary nuclear opposition. He concludes that this effect is associated with a pervasive, social psychological environment within which we share a silence on a topic that, despite the silence, still concerns us greatly.

The form of a "nuclear criticism" response—that is, a political line that does not rely on "replacement totalities"—might be suggested by Foucault's studies on sexuality and prison reform.[48] Could it be that we have been talking about nukes constantly? The film *Atomic Cafe* exposed just this compulsion, making our nuke fixation cute and archaic, but not missing the point that it was pervasive. Clearly, that compulsion survived the fifties and sixties surveyed by the film. We evoke the nuke, all the time. As Derrida and Baudrillard note, we appropriate its metaphors constantly, without irony. The terms reaction, critical mass, criticality, fallout, and disarm are a few examples. Foucault showed that the talk of sexuality corresponded with a set of discursive practices that *produced* subjects, rather than only repressing some subject that existed before discourse. Again, there are parallels in nuclear discourse. We reaffirm management, even if objectives are paradoxical. We reenact rituals of submission and fatalism. We can now even imagine such a thing as "nuclear therapy."[49] At times, the issue of teaching nukes has been nearly as controversial as that of teaching sex.

Although a serious survey of nuclear-supported micro controls would require a project different from this one, it should suffice to note that the outrageous "duck and cover" training film reintroduced by *Atomic Cafe* was unusual in one crucial way. The animated turtle who offered absurd personal strategies for defending oneself against nuclear effects was atypical only in that he so easily drew attention and ridicule. The discipline and control he advocated was in all other respects characteristic of our times. In practice, the turtle's program became part of a control discourse that had consequences not nearly as easily ridiculed.

As practiced in some parts of the United States, the techniques of nuclear survival drills were pointedly (pointlessly) distinguished from those of other disaster drills (such as tornado or hurricane drills). That is, distinctive postures, positionings, and alarms differentiated the nuke drill from the tornado drill, although there was no rational explanation for the difference. The juxtaposition of a "natural disaster" drill with a "defense" drill tends to conflate the two in a manner Baudrillard describes as distinctly contemporary; the pairing reminds us that "every event becomes catastrophe, becomes an event pure and without consequence."[50] The reality of nuclear threat (and the possibility of survival, if one took the threat seriously and obeyed one's teachers) was confirmed—not refuted—by these juxtaposed practice sessions. These drills, purporting to further chances for survival, were actually exercises in the utility of terror; only habits of hair-trigger reaction to the nuke's flash could prevent injury. We honed that reaction on a pervasive respect for military capabilities inextricably combined with an awesome fear and an empathy with Hiroshima's fate. The drills were productive, creating citizens ready to dive under the desk "in a flash" but also producing a set of fascinations. We were ready to manipulate that which sent us scurrying; post-Sputnik educational policy made the high school physics lab a well-equipped, popular place to be.

Students internalized this calibrated, produced fascination with modernity, adopting the conditions of control as their own "motives" (to attend college and study science, for some; for others, to enter the military and go to Vietnam). The turtle tutorial soon appeared old-fashioned. Coinciding with other, long-standing features that depoliticize Western society, "unspeakability" had already become an adjunct to discipline. Internalized, then labeled as repressed and unspeakable, nuclear-inspired micro controls quickly became invisible, completing the turtle's task. We ducked, and it covered.

## Language/Discipline/Politics

As I suggested earlier, nuclear criticism presents a sort of solution for the language-and-politics position as well as for opponents of nukes. Critics accuse deconstruction of falling into a passive nihilism in its avoidance of substance and essence. Edward Said, for example, has criticized this "retreat from [deconstruction's] constituency, the citizens of modern society" in the strongest of terms, but from a position within literary criticism.[51] Nuclear criticism pushes poststructuralism away from its depoliticizing tendency (if that ever was a tendency) while it disproves the charge. The rebuttal is of a congenial form. Lodged firmly within a liberal discourse, nuclear opponents have seldom proceeded beyond a value-based rejection. In the form of "nuclear criticism," opponents can appropriate a genre specifically appropriate to this task, rejecting universalisms in a way that does not subsequently reconfirm an alternate universalism. This is precisely the language-and-politics project. And since the nuke's "universalism" is explicitly textual, the approaches of literary criticism are precisely appropriate.

While the notions of survival and apocalypse push nuclear opponents in the direction of universalistic, liberal social theory, there are good reasons to reject that perspective. Poststructuralism is more than another take on domination; it is also a position on how that domination can be opposed and what roles its critics can credibly adopt. The special role of nuclear criticism, echoing Said's remarks on politics and broadening them to yet more aspects of public life, is to infuse the various poststructuralist positions with a concern for power. Whatever its faults, nuclear criticism will not leave its practitioners indifferent to the world and to their role as a threat to power, which is how Said has characterized the politics of literary criticism.[52]

From the nuclear critic's perspective, nuclear opponents unnecessarily constrain themselves when they adopt a first political assumption that is misleading and has not been subjected to the same reflection reserved for the other terms of the debate. The claim that nukes are "unspeakable" appears as an intolerable oppression, a paternalistic admonition to "be quiet" that outdistances (and silences) all such preceding calls. In carrying as well the injunction that we not speak about that which *can only be spoken*, this presumption deepens the paradoxes of nukes. There must be a reason we "cannot speak," and there is. To speak is to reveal a paradox nuclear politics must repress.

We can begin to see the awkward, reversed influence of this repression when we consider the one activity by which nuclear opponents have been attentive to language—their focus on interpretations of "nukespeak."[53] Nukespeak actually stands for a simple critique of euphemism; the problem is that this simplicity leads away from the rest of the critique offered by opponents—it is inverted. The nukespeak argument imagines a world, just before this one, which had no technocratic euphemisms and no evasive verbs and nouns. Somehow, this was a world without language games, without genres of political speech, where words "meant" something explicit. That world—presumably the one that Einstein argued the nuke inalterably changed—sits invitingly just outside our grasp, recoverable if only we would resolutely call things by their *real* names.

The nukespeak argument conforms to Baudrillard's "nostalgia for the real." There can be no reconstructive politics based on eradicating euphemism. The artifacts and practices of nuclearism have no honest, real meanings. There is no straightforward name for a protective reaction strike. To call it a bombing run trivializes it. To call it genocide (or ecocide) conflates it with phenomena—such as the Holocaust—that we would rather keep separate, for good reasons. And so on. The activities of nuclearism are encased in a context so thoroughly bound to its discourse—technologized, simulated, and then turned into spectacle—that any attempt to speak simple truth could only be another part of the simulation. This is where nuclear criticism draws the line between itself and the liberal critics of modern society, from Orwell on to the present. Either Einstein was right and the world was irretrievably changed, or he was wrong and it can be returned to a prior condition of innocence, where things had "real" names.

While the nukespeak argument suggests that these things are nameable, the nuclear critic responds that the nuke has already been actively reshaping situations. Indeed, that insight on the activism of nukespeak has been available for some time now. It was Marcuse who showed how contemporary nuclear discourse is not simple misleading, but is a manipulation that has its own syntax and grammar. Nuclear criticism might disagree with Marcuse's (and, later and more explicitly, Habermas's) presumption that a clearer language game would be closer to truth. Still, we can note that Marcuse identified the discursive component of nuclearism's authority. The forms Marcuse finds all share a pattern, constructing an inverted universe in which absurdity and danger serve security and trust.

In one important example, Marcuse deconstructs Edward Teller's oftrepeated, unofficial title, "the father of the H-Bomb," showing how this highly improbable combination is powerfully useful for discipline:

> Terms designating quite different spheres or qualities are forced together
> into a solid, overpowering whole. The effect is . . . a magical and
> hypnotic one—the projection of images which convey irresistible unity,
> harmony of contradictions. Thus the loved and feared Father, the
> spender of life, generates the H-bomb for the annihilation of life.[54]

The cleverness (or, for Marcuse, the cunning) of such formulations belies the hope that clarity could solve matters. The simplest nukespeak commentaries ask the wrong question when they argue that matters would be more obvious if terminology were less euphemistic. At a certain point—in utopian remove from the realm of knowledge and power—that would be the case. But the better question is slightly different. We could ask, What sort of an authoritative logic would behave this way? Founding strength on deterrence, then announcing the unspeakability of its topic, then producing a discourse that flaunts contradiction and turns it into unity and safety, nukespeak is more than euphemism. It turns ecstatic.

An oppositional politics, fully capable of problematizing this (hyper-) exuberant nuclearism, is possible on bases other than such suspect categories as euphemism, survival, unspeakability, and numbing. Throughout this book, I am trying to reposition antinuclearism within such a defensible political practice. At the very least, this implies an intellectual project; to paraphrase Foucault, there is a struggle over issues of knowledge, set off by nuclear criticism.

The political mood of the language-and-politics position is well framed by nuclear criticism. More precisely, a political mood could yet form, one that would contrast sharply with an existing nuclear opposition that, in the United States, has adopted a paradoxical structure, as if driven to mirror the paradoxes of nukes themselves. Antinuke talk has been ponderous—so responsible and serious that it just obviously defeats itself, and must invent the defense that "people really don't like to talk about nuclear war very much." Paradoxically, opponents then test that humorlessness by asking citizens to become independent entrepreneurs of risk, weighing the likelihood and amplitude of possible disasters. It should not be surprising that such a politics works only intermittently, if at all.

To summarize: as obvious a goal as "survival" may be, it nonetheless carries with it a series of codes and a rhetoric. Survival implies a global, unquestionable project—a faith, really—and it therefore brings along baggage we might not wish to carry. Following Foucault's model of the specific intellectual, intervening in the relations of power and knowledge, we can identify some of this baggage.

When we approach survival (and humanism, and liberalism in general) from that angle, we see some primary terms becoming far more problematic than we may have understood. The unspeakability of nukes—part of a characteristic liberal injunction to *speak*—turns out, instead, to point to a problem with the whole scheme of representation. Furthermore, our concern with technological dependence and accidents turns out to beg important issues of agency. In the wake of these discoveries, we should at least suspect that it is disciplinary power—more than technology or reticence to speak or a too-awesome topic—that has been accumulating. And in the face of that accumulation, the injunction to aid survival and counter unspeakability by simply canceling euphemism is obviously just too limited a response.

In upcoming chapters, I will try to suggest a different sort of opposition, informed by the theoretical considerations outlined above. Even if principled renunciations of the nuke—in the name of humanity or survival—have misfired, other interventions may be possible, may even be "better."

# 2

# No More Warriors

The events of 1989 — Tiananmen, the Berlin Wall, the release of Mandela — were stunning. Even so, these events have yet to produce even a ripple in the political thought of an age that is, at the same time, so obviously displaced in the wake of the sudden end of the Cold War. Each successive event was stunning all over again, but each was also transparent. Once the impossible happened, we were all left to nod at a certain inevitability, never visible before it happened.

Political analysts were only able to note a detail we might have missed, here and there, in what everyone seemed to agree was the self-evident triumph of international trade.[1] One of the things missed was this: the left was correct — obviously — all along about the weakness of the Soviets and the exaggeration of the threat they posed in Europe. The appeal of military might and huge, hyperreal weapons budgets — always a mighty drumbeat in U.S. politics — had simply collapsed, spent without results, no longer justifiable or even debatable. Obviously.[2]

This disappearance could be read several ways. Surely mainstream liberals will find a vindication of the value of peacefulness. Even at this late date, some will probably suggest that "progress" is finally starting. More cynical commentators will continue to promise conflict, now claiming that the disappearance of the military justification is itself an apparition. I would read the disappearance differently. Despite liberal warnings about how we have trained ourselves, despite conservative opportunisms, the crusty archetypes of geopolitics — realpolitik, as it is arrogantly, or perhaps optimistically, called — all received a jolt in 1989–90. Why not

see this as a harbinger of the disappearance of all sorts of hallowed fictions? If the Cold Warrior can disappear from the scene, isn't anything
possible? And, more important, how would one live, if these absences
proved durable—which is to say, if the political world really were radically negotiable?

Do not underestimate how surprising the disappearance of the warrior
was. Only months before it happened, every analyst would have confirmed that even this peacetime society—which hadn't seen war in a
generation—had found itself amidst a renewed era of the warrior. In popular culture, the Reagan era was also Rambo's time. Many of the most
popular recent films have been odes to warriors; the Rambo movies, *Top
Gun*, *Platoon*, *Hanoi Hilton*, *Full Metal Jacket*, and *Hamburger Hill*
were all successful in the late 1980s, which even saw the re-release of
*Apocalypse Now*. Warrior motifs carried beyond the toy shop and Satur-
day morning cartoon, even becoming a fashion design. For the popular
rap group Public Enemy, warrior trappings attach to radical, oppositional politics, making for a highly symbolized use of the "image" invented (and put to serious political use) a generation earlier by the Black
Panthers. After seeming to disappear in preceding decades, the warrior—
the archetypal heroic combatant—was firmly with us in the late 1980s.

Amidst this flourish of activity, the commonplace critique sees the
warrior archetype as debilitating or misleading—as propaganda. The argument I want to pose is somewhat different. My claim is that the warrior is now simply and starkly absent. For several reasons (technological,
political, and theoretical), the warrior has ceased to hold any kind of possibility. Instances where the warrior seems to be present—Panama, Liberia,
Grenada, Afghanistan, even the Persian Gulf—quickly present themselves as
failures, spectacles, or exercises in nostalgia.

At first glance, these two arguments may seem to converge. The liberal
critique frequently depends on the judgment that the warrior is "old-
fashioned," passé, or simply handicapped in view of what we now understand about human action and interpersonal relationships. The problem is that this approach ties the liberal and the militarist together, in an
opposition within which each position is defined, in large part, by what it
opposes. That mutuality has a certain resonance with the history of the
warrior. Nonetheless, I am claiming that this opposition—warrior vs. liberal humanist—not only is misleading, but is no longer possible. Such
classic juxtapositions emerge from the conviction that great values are
the context for political choice, and that this choosing is in turn at the
core of political and social life. For selection (and mature subjects) to exist, the warrior has to exist, in order to produce the grand opposition

that, in turn, drives progress. Whether mythic or historical or sociologi-cal, the warrior must be present. He[3] is the possibility (for militarists, Hobbesians, or Marxists) or the amalgam of opponents (for liberals or humanists).

## The Absent Warrior

Michel Foucault was the first poststructuralist to suggest a different ap-proach, even though he did not write about warriors *per se*. By focusing instead on the practices that now bring *soldiers* into existence, Foucault made the point that discipline is prior to essence. Discipline had a locus, the barracks, which serves as Foucault's first, key example (in *Discipline and Punish*) of contemporary discipline outside the panoptic prison. The barracks were a crucial site, a point at which postmodern discipline had accumulated, even while that development remained unidentified. At the barracks, a characteristic project—a design—emerges, matching other modernist projects. At this site, the barracks is part of the process that transforms the warrior into a modern (-ist) soldier, eventually both re-flecting and promoting his disappearance.

Foucault posed this transformation as a change impressed on the body of the fighter. He dated the change from the eighteenth century, when other transformations were influencing prisons, hospitals, and schools. Before that era, in the early seventeenth century,

> the soldier was someone who could be recognized from afar; he bore certain signs: the natural signs of his strength and his courage, the marks, too, of his pride; his body was the blazon of his strength and valour; and although it is true that he had to learn the profession of arms little by little—generally in actual fighting—movements like marching and attitudes like the bearing of the head belonged for the most part to a bodily rhetoric of honour.[4]

By the eighteenth century, "the soldier has become something that can be made." The conscription agent no longer looks for someone who could be a soldier. The training officer will take care of that; "out of a formless clay, an inapt body, the machine required can be constructed." By inscrib-ing small increments of discipline onto the body, training transforms these recruits. "Posture is gradually corrected; a calculated constraint runs slowly through each part of the body, mastering it, making it pliable, ready at all times, turning silently into . . . habit."[5]

Foucault closely examined the new techniques through which this change occured, finding in them an example of a special (if generally un-noticed) change in the character of power in Western societies. Indeed, the soldier's alteration provided Foucault with his central image of "discipline." While scholars were researching such primal models as "the warrior," the disciplinary power described by Foucault was already creating something new. "While jurists or philosophers were seeking in the past a primal model for the construction or reconstruction of the social body, the soldiers and with them the technicians of discipline were elaborating procedures for the individual and collective coercion of bodies."[6]

This is such a profound transformation that it promises more than a changed fighter, and something other than merely a critique. This transformation pursues two directions. Looking for grand, metaphysical schemes for understanding society, we find the "lowly," particularistic organizing themes of discipline and surveillance. In the other direction, looking for those archetypal individuals who would confirm or confront those grand schemes—the king, the legislator, the warrior—we find instead bodies and selves in constant change. In each case the pattern is the same. What we had thought was clear presence turns out to be difficult to locate. The shifting character of language moves to the foreground, while the seemingly solid nature of what language talks about (everything) becomes vague. A postmodern theme emerges: the proliferation of empty signs—a generalized, but still unexpected absence.

Gilles Deleuze and Félix Guattari make political use of that sense of absence, showing that actual warriors exhibited "fundamental indiscipline, . . . a questioning of hierarchy, perpetual blackmailing by abandonment or betrayal, and a very volatile sense of honor, all of which . . . impedes the formation of the State."[7] Thus, on one hand, successful creation of the state mandates a new kind of citizen, crucially including a new army. On the other hand, the sort of war emerging with that new state changes fundamentally, robbing the warrior of his role and site. The knight begins to receive a wage and will not be a knight much longer. Then, as Deleuze and Guattari note, the emergence of guerrilla war signals further absence. "War in the strict sense . . . does seem to have the battle as its object," and the battle is the warrior's supreme site. But guerrilla warfare "explicitly aims for non-battle."[8] Before such warfare emerged, the warrior had always presumed the battle; the two were inextricably connected.

In another sense, this notion of de-territoried guerrilla war is only the sign of a greater absence. We already know that war itself—not just a particular variety of war—exhibits the effects of this transformation. We can begin tracing the process of change with Paul Virilio's claim that the technology of contemporary war amounts to a new kind, not just a new means, of war. The inherent goals of the war machine have changed. It is not only the destructive potential of the bomb that triggers this shift. The submarine, bomber, fighter aircraft, and missile join the bomb, ushering in a new vision of war. Virilio argues that speed and movement replace territory as the keys to this new war. Territory disappears when an ICBM can flash overhead, instantly outmoding the complex logistical concerns that had hounded all previous armies.[9] But, Virilio notes, even earlier technologies had promised this demise of territory as the central object of war. Tanks—which Virilio calls sans (rather than all) terrain vehicles—extend war "over an earth that disappears, crushed under the infinity of possible trajectories."[10] Terrain stops neither tank nor missile, and the object of war machines becomes pace.

However thoroughly Virilio traces the transformations of *things* into *movement* (velocity, speed), he may not fully recognize how thoroughly this transformation becomes disappearance, de-formation.[11] The movement he finds at the center of political phenomena (not only war, but revolution and other events as well) is not a new imperative so much as it is a sign that immediately cancels itself. Virilio analyzes events and techniques in terms of their goals, their targets. The eventual consequence of all this displacement (of a "metaphysics" of things transmuting into a teleology of motion) is a compounded absence. Movement wipes out terrain, only to be displaced, in turn, by deterrence. The warrior has begun to be displaced (even before the advent of nuclear weapons) by techniques of barracks and training field discipline. That displacement compounds with development of the tank, then compounds again with the development of aircraft, then again with missiles. When the missile obtains a nuclear warhead, all this motion stops, in a sense. What is left, the residue, is the nothingness beckoned by nuclear war. And that goal, in turn, is quickly displaced (in a shock of horrified recognition) by the mere threat of nothingness. Teleology ends in deterrence, in the commitment to disallow the movement of events to their designed ends.

In his *Critique of Cynical Reason*, Peter Sloterdijk better situates the demise of the warrior, identifying it as a major moment in the accumulating cynical age. He traces (at least a) double disappearance of the warrior. First, the warrior is denaturalized by the requirements of deterrence: "All modern military ethics . . . have abolished the image of the aggressive hero because it would interfere with the defensive justification for war."[12] Second, following that political development, technology finishes the task of displacing the heroic subject: "There are modern artillery systems that in strategic jargon are called 'intelligent munitions' or 'smart missiles,' that is, rockets that perform classic thought functions (perception, decision making) in flight and behave 'subjectively' toward the enemy target." The "human factor" present with "the self-sacrificing kamikaze pilots" or the manned bomber "is fully eliminated." This is an elimination of special status: "With the 'thinking missile,' we reach the final station of the modern displacement of the subject." In the cynical age, the "human body in the society of labor and war had already long

been an artificial limb even before one had to replace damaged parts with
functioning parts."[13]

Avital Ronell has developed yet another strategy for dislocating the
warrior, situating that disappearance in a more generalized absence.
More than the warrior disappears in the nuclear age; the body in general
is displaced.

> Arms have become arms that race, erasing the bodies' members,
> heads turned into war heads, and so on. By several rhetorical
> maneuvers, then, the body has already been evacuated from the site of
> battle, a battleground no longer being grounded as a circumscribable
> place where a class of warriors might engage one another in a limited,
> classical way.[14]

Virilio extends the point in *War and Cinema*, emphasizing the filmic form
of new weapons guidance systems. "The fusion is complete, the confusion perfect: nothing now distinguishes the functions of the weapon and
the eye; the projectile's image and the image's projectile form a single
composite. . . . The old ballistic projection has been succeeded by the projection of light, of the electronic eye of the guided or 'video' missile."[15] In
the next chapter, I will examine that "evacuation" further, arguing that
the robot has begun to function in several rhetorical ways in nuclear discourse. For now, Ronell, Sloterdijk, and Virilio combine to show, from
different angles, how difficult it will be to establish "an angle" from
which to view, much less resuscitate, the warrior. The governing sign is no
sign at all—absence now rules the possibility of the warrior.

It is worth noting, as an aside, how this analysis of disappearance influences a well established area of antimilitary concern. We can now
understand why years of protests against "war toys" seemed largely unsuccessful. Although there is little to suggest that the shift was conscious,
it is clear that opponents finally began to make progress when they quit
treating those toy guns, for example, as real, and began treating them as
dangerous fakes. When critics argued that the guns produced warriors,
the argument failed; at some level, we knew you could not make warriors
anymore, and that new disciplinary methods could create soldiers,
whether or not the recruits had childhood histories of having played with
toy soldiers and water-filled machine guns. The soldier would be made by
marching, tightening covers on bunk beds, and shining shoes, not by manipulating simulated warrior icons. Replacing this false sense of training
with the understanding that the toy gun was a simulation which could
have consequences (it could be confused with a real gun), the opposition
to these toys made at least some gains.

To summarize: the warrior disappears first when disciplinary power
transforms fighting bodies. The guerrilla's battle-evading tactics spread a

sense of frustration over the warrior's characteristic site, the battlefield. Then the warrior's goal—the capture of territory—is canceled. Then terra itself is left behind, to the airspace of planes. Then, beyond air and (merely geometric) planes, the missile adds another layer, beneath which the memory of warriors could be said to be buried. Finally, a logic of deterrence and multiple technologies associated with the demise of the subject emerges, produced at the end of this chain of transformations. "Deterrence," then, functions as a sign of a special type, pointing to multiple disappearances but not to any force or condition which would replace the absent warrior.

## The Contrary Mission

The "warrior" (so distant, so absent that his category must be bracketed) acquires the strangest mission of all. His orders are to prepare to fight so that he will never fight. This odd preparation—training to fight very well, so that one will never fight—is different from the old pose of strength adopted by all previous warriors. Not only did those warriors *expect* to fight, sooner or later; they did not know the possibility that they would *not* fight. Every army and almost every generation of warriors had awaited their chance, and then had fought. We non-warriors ("civilians") should not have expected today's "warrior" to move easily into this new era. The frustration that marks these soldiers should not surprise us. Contradictions, ironies, and absurdities had begun to surround them in ways that mere enemy warriors never could.

In one example of the twists a modern "warrior" encounters, Seymour Hersh describes Major General James C. Pfautz, one of the intelligence figures central to the aftermath of Korean Air Lines Flight 007's downing in 1983. In the view of some of his subordinates, Pfautz was a "martinet," a committed Air Force man, perhaps even excessively partial to disciplinary rigor. Still, this thoroughly military man could sometimes defy convention.

> As a student at the National War College in the early 1970s, he had earned a reputation as someone who was 'that way'—soft on communism—because he had refused to solve all problems in war games by calling for a nuclear strike. "Everyone's solution to a problem was 'Nuke 'em,' " Pfautz said, "and I'd object and say 'Let's try to work it out diplomatically and politically.' "[16]

The contrary mission shows up, continually. The War College simulates conflict that concludes with simulated Nukes. With the advantage of hindsight (but the disadvantage of ideological militarism and antimilitarism), we might not expect Pfautz—the rule-bound militarist who rose high in the Pentagon bureaucracy—to have been the one to propose a different (and more peaceful) endgame. But that proposal was the sanctioned one; if deterrence works, even war games are resolved by negotiation. However, when Pfautz suggested such a response, he was seen as "that way." The absurdity of these combinations is then compounded when Pfautz inadvertently plays a central role in the simulated, fabricated orchestration of a major powers tiff over KAL 007.

Michael McCanles's close textual analysis of the paradoxes of the deterrence discourse shows how the counterbalance of nuclear threat is "theoretically and practically unstable and undependable."[17] McCanles articulates the paradoxes of the deterrence discourse by noting how the characteristic statements about deterrence contradict the larger (textual) contexts in which they reside. Citing examples from Brezhnev and Weinberger, McCanles shows how pervasive the paradox is. The Secretary of the Air Force, Verne Orr, put it most simply in a 1982 speech to the Los Angeles World Affairs Council:

> We have to be stronger than we were before because the Soviets are stronger. Otherwise we will not be able to deter the conflict we all want to avoid. And if deterrence does fail, we must be able to win to survive.[18]

Preparing simultaneously to "deter" and to "win" is paradoxical; these ambitions belong to different, mutually exclusive discourses. The paradox remains at equilibrium only if both sides repress any discussion of the obvious disequilibrium their statements imply. Amidst this repression, contradiction, irony, and absurdity, the supposedly central character—the warrior—has slipped away, unnoticed. These are the signs of absence.

Starting to look for the warrior, we inevitably find deterrence. In the language of deconstruction, the warrior is a floating signifier, a sign that refers to nothing. It is not just that the warrior is outmoded or old fashioned, but that he has vanished. This disappearance is of interest precisely because our recognition of it leaves such a gap (for Derrida, an *aporia*) between our talk and our understanding. To discuss the "meaning of the warrior," we will have to investigate the errors or misplacements that surround this missing referent, more than the qualities or characteristics that would confirm his presence. This shift is not abstractly intellectual; it has huge consequences for militarists and their opponents, as well as for every political theory or ideology. If the warrior seems omnipresent at exactly the moment when the possibility of his existence has vanished, the issue of reality itself is at stake.

The 1980s television advertisement for the U.S. Marines, in which the "making of a Marine" parallels the manufacture of a sword, teemed with ironies. These few brief seconds of video indicated a shift, even if they did so in an inverted manner. Marines *are* made, and their manufacture has traced a path not unlike the production of their tools. The craftsmanship of the sword maker is a focus of a nostalgic hope—the almost desperate quest to retain a sense of honor and chivalry even though those attributes have very little to do with modern, post–Khe San and Beirut Marines. The ceremonial marching the full-dress Marine stands for is one of the practices marking his difference from the medieval swordsman. Bayonets still exist, but they are manufactured (not crafted), and are not ornamented. If the Marine actually uses his bayonet, it usually means that guerrilla "battle" has been engaged, and any such occurrence, now, must necessarily remind us of other recent times when the warrior Marines have tried (usually in vain) to confront an enemy in battle.

This is what Timothy Luke has recently called a "teletradition," a cynical, instant, powerful inversion.[19] In the ecstatic simulation of the warrior that now permeates popular culture, Marines can even respond to absence. The Corps, as they call themselves (core and corpse), advertises for recruits by openly inviting them to be the raw material for a manufacturing process. They "need a few good men," just as the munitions factory needs a few good casings. The phrase is rhetorically sophisticated. It understates ("a few") as a reminder of elite ritual. Promising symbols of honor when none can be delivered, the Corps's ad agency has not deceived so much as it has admitted the Corps's role in a simulation, turning this concession into evidence that Marines can even deal with a postmodern world that overruns the symbols that had guarded their perimeter.

The U.S. Armed Services relied for generations on an "economy of obligation." All males were understood as "obligated" to do military service, even if no strict legal compunction applied. Even after Americans stopped feeling this obligation to serve, one could still choose to enlist for a strenuous and lengthy commitment to a more honorable branch of the service (the Marines), which would in turn avoid (if possible) using recruits rounded up by the Selective Service. Thus, reserving honor as an economic good available to be earned, the Corps had established a ritualized way of guaranteeing the value of the "contract" it offered recruits.[20]

The payoff on promised honor has always required action on a battlefield. In his essay on Clausewitz, Garry Wills defines the source of that honor as a prenuclear reversal essential to the warrior's code.

Hatred assimilates each side to the other, the *Wechselwirkung* [a reciprocal altering, one of the other] becomes an oscillation as well as a mutuality, each side *becoming* the other as the sides work each other up to a resemblance in extremism. There is even an altruism of hatred, in which the other becomes more important than the self. One is willing to

give one's own life in order to take the foe's. This is the selflessness that
gives war its weird and inverted nobility. It is, in Clausewitz's view, a
necessary mark of fighting men.[21]

That mark becomes less and less available when rationalism takes
over, or when war is managed and limited, or when the foe is a guerrilla.
Thus, the bitterness and revisionism of U. S. veterans of the Vietnam War
is more a response to the demise of the warrior archetype than to any
actual affront encountered after the war ended. The oft-repeated
misdiagnosis—"We could have won if the civilians had let us"—turns out
to be correct, if still misplaced. Under managerial command, pursuing
limited symbolic goals (as opposed to the unlimited ones of a past era,
evoked by Wills), these proto-warriors—out to experience their craft in
the field—found, instead, their absence. The issue of stateside parades is
irrelevant in the face of the actual, twisted source of honor whose site
(also sight, and cite) has always been the battlefield.

Seen in this grid of political control, deterrence, and technologized
pace, the plight of American veterans of Vietnam is easier to understand.
All wars have featured the political "interference" that so bothered Vietnam-era "warriors"; this time, however, that interference represented the
future absence of such byplays between warriors and politicians. Warriors have always pined for the (dangerous, self-obliterating) battlefield,
where they were "most real." But the Vietnam veteran's pining has a different, more final quality. It isn't just *he* who has lost the battlefield, it is
also his class, the tradition in which he has tried (futilely) to lodge himself.

It is a classic Foucauldian case; the warrior was produced by an economy impressed on bodies, an economy that created calibrated, incremental outcomes that could be grafted onto an ideology of "choice," at the
same time displacing other concerns and desires. This is a broader "production" than the one imagined by contemporary critics of the warrior.
Even the madness of battle could be made useful. But in that process, another transition became possible. The classic archetype remained only as
residue, as nostalgia and complaint; the genealogy of the warrior now
contained a gap.

It is tempting to write the warrior off, to giggle at the hugely hilarious
Rambo and to see the contemporary soldier as the true Maytag repairman, always ready but never called.[22] Fully competent but fully inactive,
the Maytag advertising symbol and today's putative warrior do share
characteristics. Visible, successful, able, and presumably useless, they are
brothers. But our sarcastic laughter returns to us, with a bite. When such
odd, misplaced political rhetoric seems to dominate the discussion of
such an obviously crucial question, we accurately call the situation "dangerous."[23] It could be that this perceived danger has informed the now

famous flight of postmodernists from (what others call) real politics. To best assess the politics of this "flight from the real," we should consult the central theorist of that flight.

## Demise of the Real

Jean Baudrillard—the analyst of the disappearance of the real—argues that the real has dissolved into simulation and "hyperreality." This argument aims explicitly at the metaphysics of nature and essence that liberal humanism and warrior nostalgia share. Once the connection between a sign and some "thing" is convincingly severed, "facts no longer have any trajectory of their own, they arise at the intersection of the models; a single fact may even be engendered by all the models at once."[24] This disconnection is dangerous; the basis of allocation, adjudication, and all judgment becomes problematic.[25] Thus, "order always opts for the real."[26] One might presume that opposition would opt for the opposite— for the floating, giddy rootlessness that order would extinguish if it could. But the attractions of certainty and ideology are not frivolous; the real dies hard, especially when the wild destruction of contemporary war is at stake.

Baudrillard begins his discussion of nuclearism with a characteristically slippery metaphor; the central effect of nuclear technology directs itself inward, absorbing any outward-directed or instrumental impulse. The bomb, and its key mode, deterrence, "leukemizes."

> It isn't that the direct menace of atomic destruction paralyses our lives.
> It is rather that deterrence leukemises us. And this deterrence come[s]
> from the very situation which excludes the real atomic clash—excludes it
> beforehand. . . . Everybody pretends to believe in the reality of this
> menace (one understands it from the military point of view, the whole
> seriousness of their exercise, and the discourse of their "strategy," is at
> stake): but there are precisely no strategic stakes at this level, and the
> whole originality of the situation lies in the improbability of
> destruction.[27]

As Baudrillard surely intended, the leukemia metaphor is not clear; it could be read several ways. One of its aspects is, of course, that nuclear fallout can cause leukemia. In the quest for complete control, nuclear technology induces (utterly inevitable and tragic) "side effects." Paralysis is a metaphor familiar to Americans old enough to remember the polio epidemic of the 1950s, but leukemization is more to the point of our present dilemma. This was, after all, the first cancer to have its own name: cancer of youth, of blood (hence, of flow).

The loss of the real, too, is an internal debilitation, one that leaves us to function with seeming normality, but only for a while. The invisible process of the real's demise parallels the invisibility of fallout itself. In turn, fallout is a metaphor for unintended, unanticipated consequence; we speak of "the fallout from this deal." For Baudrillard, "the true nuclear fallout" is that the spread of the social is now governed by the pursuit of "maximal security and deterrence . . . The meticulous operation of technology serves as a model for the meticulous operation of the social. . . . Nothing will be left to chance." In American slang, we would say that society "had a falling out" with the very concept of society. Trained by expansive technocratic claims to expect that this socialization process would be explosive (that is, revolutionary), we are understandably reluctant to see it as an "irreversible, implosive process, a generalised deterrence of every chance, of every accident."[28] Fallout "falls out" of the sky, uninvited and undetected, with consequences nonetheless unstoppable.

Baudrillard is no utopian, but he (like all prophets of control) must face charges that his critique is an ideal. There are, still, wars and armies, despite his claims about deterrence and disappearance. He concedes as much, addressing the question in a discussion of the Vietnam War and its earlier parallel in Algeria. Both were wars, but Baudrillard argues that they are distinctive in their shared goal, or subtext. It was the "tribal, communal, pre-capitalist" structure of the real that had to lose those wars, not one side or another. The adversaries share this ambition "behind the armed violence, the murderous antagonism between adversaries." Their armed opposition "seems a matter of life and death, and . . . is played as such (otherwise you could never send out people to get smashed up in this kind of trouble)."[29]

> The flesh suffers just the same, and the dead ex-combatants count as much there as in other wars. That objective is always amply accomplished. . . . What no longer exists is the adversity of adversaries, the reality of antagonistic causes, the ideological seriousness of war.[30]

In this way, Baudrillard can dismiss the usual, serious suspects; proliferation, escalation, and accident are each beside the point. "Responsibility, control, censorship, self-deterrence always increases faster than the forces or weapons at our disposal: this is the secret of the social order."[31]

This is Baudrillard at his most Foucauldian point; the social order issues the signs of militarism, but the signs are empty. More precisely, those signs serve as diversions from a secret. It is an open secret, to be sure; the influences of disciplinary power are out in the open, available for anyone to analyze or criticize. But under the sign of the nuke, that power operates as a secret anyway, no matter how visible it is. In short, the nuke remains a crucial example, but in a somewhat different way than we had suspected over the last half century. The nuke was the most important empty sign; it was the crucial rehearsal of the type of power Foucault identified as characteristic of our post-Oppenheimer age.[32]

As the 1990s begin, we must note that Baudrillard's broad statements about deterrence now seem outmoded. "Deterrence excludes war—the antiquated violence of expanding systems. Deterrence is the neutral, implosive violence of metastable or involving systems. There is no subject of deterrence any more, nor adversary, nor strategy—it is a planetary structure of the annihilation of stakes."[33] The generalized statement of an end to politics, read in the context of Gorbachev's realignment of Eastern Europe and new defense reductions throughout the world, surely rings false. It helps clarify Baudrillard's position to note that, while his statements have a very broad style, the question he addresses is quite specific. On the one hand, contemporary governments prefer spreading disciplinary control to conducting war. On the other hand, the destructive potential has already made such a war an impossible option; "the entire myth of the revolutionary and total strike collapses at the very moment when the means to [do] so are available—but, alas, exactly because the means to do so are available. This is deterrence in a nutshell."[34] In either case, Baudrillard's analysis conforms to the Gorbachev era, in the sense that military stakes have been overcome by commercial, communicative, and other stakes.

What is not confirmed is Baudrillard's idea that politics disappears for the same reasons that the warrior experiences a demise. At this early stage of the post–Cold War era, it seem possible that politics has been revitalized, unfrozen after the reorganization of stakes and methods. In any case, whether Baudrillard will acknowledge it or not, a political conclusion emerges; under the sign of a war that will not happen (or, if it does, will leave our conclusions unwritten, since it will end writing), a new form of authority consolidates its position. But new forms of political response and opposition may also be enabled.

## Nuclearism/Humanism/Foucault

In a sudden and recent development, antinuclear politics may have begun to take deterrence quite seriously. In one important example, Randall Forsberg (founder of the nuclear freeze movement) for a long time pursued the freeze as an intermediate goal, but always understood that disarmament was the only acceptable ultimate goal. Forsberg nonetheless found in the 1987 Washington summit an opportunity to change positions. In doing so, she reaffirmed a set of distinctions that become more important as we enter an era of arms reduction. Within the approach I

am pursuing, we could read Forsberg's reassessment as the next ground
on which liberal humanism would respond to nuclear criticism.

> Nuclear weapons are teaching us how not to be violent. We have learned
> that you don't use everything you have. You choose not to. It's only in
> the nuclear era we have begun to practice this, not just in terms of
> nuclear weapons but in conventional war. . . . The ability to be violent
> exists in all of us, but it isn't uncontrollable, inevitable.[35]

This is "Just Say No" for antinuclearists, another hopeful wish that
choice and will could prevail, even though the same wish has failed completely, time and again. What is more important about Forsberg's assessment is how closely it duplicates a pattern Foucault found, especially in
the institutions of criminal justice. Under the sign of humanistic progress,
subtle patterns of discipline and surveillance flourish. In Forsberg's telling formulation, we are "taught"—the classic metaphorical mechanism
of humanistic politics—but the "teacher" is now the (inert, invisible, but
symbolically overdetermined) nuke. Because the ostensible "policy area"
seems so important, the seemingly secondary, diffused concerns about
control recede, replaced by an ever more fantastical pedagogy.

Foucault suggested that the seeming "humanization" of criminal justice practices actually contributed to a different project, more important
than a mere choice of policy agenda items. Indeed, the distinction points
toward a fundamental rent in the philosophical fabric of our political life.
For humanists, human nature is the crucial political factor. Political structures (including our understandings of concepts, institutions, and events)
spring from a modernist, liberal notion of that nature. Even presumptive
political opponents, as the Forsberg passage shows, are motivated in the
context of a shared nature. At some level, the humanist argues, the violence inherent in a national foreign policy or weapons strategy implicates
all of us, reflecting that shared nature.

Humanist progress thus requires, most of all, success in harnessing
that human nature, in retraining it so that we "choose not" to be as vio-
lent as our nature pushes us to be, and as our inventiveness with tools
now encourages us to be. Foucault's entire political approach rests on a
reversal of this presumption. For him, this retraining project is the crucial
(i.e., theoretically privileged) political action, displacing its supposed target, human nature. The acts that harness—not the subjects of those
acts—are the characteristic moves in the politics of our age. The authority produced as a "mere vehicle" for all those liberal values becomes more
important than the message it supposedly serves. If Foucault is right, this
is a momentous reversal, an inversion of terms easily as important to us
as Marx's displacement of the individual into the social and the ideal into
the material have been for the politics of the last century.

Foucault lodged this argument in his discussion of the self, the modern creation that replaces any imaginable "human nature" with an ensemble of techniques (carried by discourses) that modern individuals use to modify themselves endlessly. The modern self was invented and refined. We become calculating, self-monitoring individuals, ready to be treated (by medical and mental health professionals), and willing to work endlessly on these new selves. This process of transformation works, in part, through every kind of public "care." The madman, who had been a sign of the demonic other, becomes a mentally ill patient, who is shaped into a contributing (read "working") member of society. To the extent we continue to act as if the madman and madwoman *themselves* were still central fascinations of social life — as indeed they once were — we have simply missed the crucial transformation that creates them and gives them a public role. Foucault traced similar transformations of the criminal as well, in terms that could apply directly to the warrior. Such an assessment would conclude that a compulsion about warriors is archaic, since it ignores the transformation that has steadily canceled the warrior. To debate whether "courage," on the one hand, or "learned nonviolence," on the other, is the crucial cultural signal, is to miss the point.

Modernity creates autonomous modern selves, but if this "freedom" is too casually glorified or assumed, we miss the implication of self-control that accompanies the transformative process. This control accumulates in diverse and multiple practices, habits, and discourses. Although the effective ordering of society is indeed at stake, no single agent (individual or class) directs this process. The attempt to analyze power in this century therefore fragments and localizes, arriving at common language but no common agent of oppression. Instead of tracing all oppressions back to the class struggle, poststructuralist analyses spin outward, identifying more and more agentless, disorganized projects that nonetheless produce effects of power.

In nuclearism, two related effects gather beneath the image of such seemingly progressive goals as arms control. First, the focus on survival that informs antinuclearism reflects a narrow view of life, overshadowing almost any potential reconstruction of the processes that actually constitute that life. To rephrase another Foucauldian position, nuclear technology does not merely repress life and negate prospects; it forms practices that actively shape what that life will be. The second effect follows from the first. Under the sign of "survival," all sorts of surveillance, discipline, and covert operations can positively act to reshape that self (and along with it, social life). Unless we assume a psychological (rather than political) mechanism, we cannot then say that the fear or trauma of thinking about nuclearism necessarily induces a certain paralysis, or inability to act. The reluctance to politicize problems is always already present in a society that medicalizes, psychologizes, or otherwise professionalizes every worry. Under the nuclear umbrella, that reluctance expands. The cult of the expert—always a depoliticizing ploy—accumulates when the experts can produce such extravagant displays as the mushroom cloud.

In turn, those experts remind us that new forms of surveillance are obviously "called for" by the new, common facts of our social life. "Verification"—the new euphemism for surveillance—is now unanimously endorsed. Even critics of militarism answer qualms about verification with an oblivious commitment to the hyperverification promised by efficient, omnipresent spy satellites. Rather than taking offense at such devices, arms control advocates praise them as the precondition for "trust."

When critics discover that the FBI still engages in monitoring political groups for reasons of "national security," their response usually is that such surveillance is redundant, given the obviously moderate nature of its targets. When press reports revealed (in early 1988) that the FBI had collected information on hundreds of individuals and organizations opposed to U.S. policy in Central America, those reports focused on groups such as Atlanta's Southern Christian Leadership Conference, Chicago's Maryknoll Sisters, and the United Steelworkers Union. Implicit in that approach is the judgment that these are hardly worthy targets of surveillance. The radicals have already trained themselves to follow a less dangerous path, even while sometimes claiming that their new (therapeutic or professional or even entrepreneurial) discourses are still radical. There is another implication of this story. Such "surplus surveillance" reminds us that the spying itself is the point, regardless of any danger its targets might pose.

In short, the spy (the self-described "spook"), not the warrior, captures this era. Oliver North gained prominence by standing ("at attention") between these models, hence offering something for everyone.[36] But even if all the stylized warrior mannerisms were perfectly realized, no Lieutenant Colonel could now become so visible on warrior merits. North was a covert actor, willing to "take the rap" only after he'd been uncovered. That his covert activity pursued largely symbolic goals—the continuing U.S. simulation of geopolitics in Central America—only adds to his celebrity.

Each of these instances—North, arms control debate, the reification of both survival and surveillance, and all other events of the nuclear age—issues from a crucial absence, more than some material condition or theoretical structure. The warrior's absence is no abstraction; it highlights a role of absence that becomes visible throughout all discussions of contemporary war and peace. The central fact of the bomb is its lack of *use-value*. This is James Der Derian's claim in his poststructuralist reading of deterrence's implementation in *On Diplomacy*. Der Derian argues that "nuclear weaponry has corresponded to the emergence of a techno-diplo-macy which relies on a balance of terror and mutual deterrence." This represents a new diplomatic era in which "two paradigms of diplomacy have continued to coexist," with non-nuclear weapons balancing power while the "*non-use* of nuclear weapons [balances] terror."[37]

Der Derian further emphasizes absence when he uses alienation to an-alyze the diplomatic realm. "Marx's conception of alienated labor offers [a] parallel to the effects nuclear weapons and mutual deterrence have had on both human and diplomatic relations."[38] Der Derian reminds us that Marx was working on the disjunction between signs and material use-value, and that the resolution was distinctly political;

> In Marx's scheme, the separation of an object's use-value from its exchange-value can give rise to a commodity fetishism. No object can become a commodity without first having some use-value, that is, the power to satisfy, directly or indirectly, a human need. But *qua* commodity, the object has only an exchange-value, or a power of exchanging against different quantities of commodities. Consequently, says Marx, it does not 'contain an atom of use-value'. In the context of a modern society, the result is 'a definite social relation between men, that assumes, in their eyes, the fantastic form of a relation between things.' . . .
>
>   For reasons that should be obvious, a material exchange [of nuclear weaponry] against other quantities of other commodities is not feasible. It is, rather the *threat* of an exchange which valorizes nuclear weaponry.[39]

This politicized threat (in the name of exchange) leads to a particular kind of politics—in Der Derian's judgment, to "neglected develop-ments," the most important of which is "the high priority now given to surveillance between states and within states. . . . Crypto-diplomacy is ascendant." "The Panopticon's combination of total visibility and tech-nological power has more than a metaphoric value for the *telos* of omni-science and omnipotence which is implicit in techno-diplomacy."[40]

The principle of efficient control Foucault finds epitomized by Ben-tham's Panopticon matches diplomacy in the era of the bomb. An unseen but all-powerful agent may (or may not) hide behind the blinds, watch-ing. What the (absent) agent could do is a matter for concern, but be-neath that fear another kind of power advances. The object of that gaze learns to discipline herself, whether or not the gaze is actually on her. As Der Derian notes, this approach concretizes alienation, making it less ab-stract, less of a "theorist's lament." We should also note how crucial *ab-sence* is in this relationship; it is the condition of an effective diplomacy in this era—it is the condition for the era itself.

## Replacement and Therapy

Are we compelled, then, to find something to replace the absent warrior? I have suggested that the spy might be that replacement, but the spy is another absent agent who may or may not be present, a ghost who may merely (and falsely) imply a presence. There have been other suggestions, as well. Feminists have sometimes suggested that the warrior impulse satisfies a need for sociability among men, and that a therapeutic discourse might diffuse this dangerous need.

While a full critique of the therapeutic alternative is beyond the scope of this book, I can at least note that such a solution only displaces one power grid with another. Therapists diagnose militarism—such an inexplicable attitude when it has been so completely displaced by deterrence—as a psychological residue that should be treated. That treatment raises other issues, however. As Foucault argued, such objects of therapy can turn out to be more like compulsions themselves, implying all sorts of positive agendas.[41] The relentless self-surveillance implied by therapeutic contexts does not "make us well"; it submits us (or rather, finds us submitting ourselves) to a new regime of power. The victim of such "progress" is, always, the political.

In this instance, as in every humanistic prognosis, we must ask whether this progress means a humanist development or the advance of discipline and surveillance. This is the subtext of discussions in which poststructuralists confront therapy, deterrence, peace, and other solutions, arguing that each presents a new regime of contemporary power, not some progressive outcome. A crucial project, then, would be to trace the disciplinary practices linked to nuclearism. Another, as I have already suggested, is to examine Baudrillard's radical suggestion that politics ends in this age.

The analysis of absence puts antinuclearists at a point of choice. On the one hand, deterrence (and its associated features of control, discipline, and surveillance) might still function as Baudrillard implies, hence undermining any critique of these weapons' mere existence. The other choice is to admit that the "ideological seriousness" of "the adversity of adversaries" is still functioning, despite the lack of nuclear hostilities during the Cold War and despite the onset of the Gorbachev era. Wishing to criticize both the seriousness and the control, opponents have accepted the possibility that these two poles might coexist. The emphasis on accidents might reflect a way to combine them; ideological conflict could erupt through the veil of control, given an accidental opening. More to the point, Baudrillard's analysis forces us to ask whether nuclear opponents have kept the nostalgia for warriors alive, just as militarists have. The warrior is convenient, a dodge that falsely reaffirms the continuing

seriousness of war preparations, finding seriousness in psychological archetype even when it is no longer the main feature of actual military
establishments.

As his critics have accurately noted, Baudrillard resolves this political
dilemma for himself simply by leaving no plan for establishing "critical
distance."[42] No longer can we assume that an independent vantage point
of meaning exists. Foucault articulated this loss of critical perspective by
detaching the effects of power from any agent who could meaningfully be
said to "exercise power." Baudrillard consciously goes further; we cannot
even assume the critical posture, or at least can no longer be convincing in
that pose.[43] This position thus makes it possible to articulate the warrior's absence without a hint of replacement. No alternative icon—no
peacemaker or healer—displaces him. No process—whether therapeutic
or grandly intellectual—substitutes for the processes of war. No new
villain—whether psychological nature or oppressive class or conspiring
agent—exists to justify a political critique, at least in any familiar model.
The warrior's absence is, instead, stark, simple, and complete.

It is that sort of absence—not only the demise of the metaphysical—
that makes Baudrillard a nihilist, and gives his nihilism its character. It is
the nihilism of the disappeared warrior, as well as the disappeared God.
This transparency has glossy television ads, replacement habits for the
citizenry that remains, even a rhetoric of nostalgia. What it does not have
is the darkness one looks for in a nihilism.

> Nihilism no longer has the dark, Wagnerian, Spenglerian, sooty
> complexion of the end of the century. It no longer arises from a
> Weltanschauung of decadence or from a metaphysical radicality born
> of the death of God and of all the consequences that necessarily follow
> on from this. Nihilism today is a nihilism of transparency, and in a way
> it is more radical, more crucial than in its earlier historical forms, for
> this transparency, this floating is irresolvably a floating of the system,
> and of all theory which still claims to analyze it.[44]

In Paul Foss's terms, Baudrillard is a "positive or active nihilist," one who
"becomes 'ecstatic,' he takes leave." This leave taking, Foss concludes, is
"supremely joyous and in no way despairing."[45] Others have used the
(purposefully shocking) term "intellectual terrorism" to describe the activity, but the more precise term, I think, is "intervention."[46]

The postmodern activist intervenes, launching an insistent pursuit of
the habits and practices that imply control. As I have been suggesting, the
habits, practices, and rhetorics of self-control will be a special site for this
intervention. Sad though it may be for some, there is no warrior available
to contrast with us more sensitive selves, to be our foil. And these sensitive selves are, in turn, the residue of the same kind of power that made,

then deposed, the warrior. The absence, then, of both the humanist and the warrior leaves us to expose control, without being able to promise a progressive regime, but also without despair over the absence of such a plan. To find ways to sustain that intervention—to make it lively, even attractive—is the actual task of a political postmodernism. This is, as well, the project to be undertaken at this moment of the warrior's demise.

# 3

# Robotics (The Bomb's Body)

The humanoid machine, perhaps semi-autonomous, has a long history in Western thought. As Andreas Huyssen notes, when eighteenth century scientists began to understand the human being in terms of machine metaphors, popular culture followed suit.[1] Performances of automata who sang, played musical instruments, or danced became major hits. But these embodiments were only the first, enthusiastic expressions of scientific breakthrough. As the nineteenth century brought the image of the threatening android to the fore, danger replaced hope and confidence. No subsequent embodiment of hopes could reconstruct the naive delight of the previous century. The twentieth century names those pseudopeople ("robots") and gives them a special place in both scientific and popular culture. The robot theme is found in science fiction books and movies, but also in advertisements, editorial cartoons, pop music, dance, and the visual arts. Even if the Artificial Intelligence (AI) masters took far longer than they first predicted to produce any convincing intelligence algorithms at all, they can produce enough of an image to keep this grand displacement going. Their reinvigoratation of the nineteenth-century automata finally completes the Cartesian dualism. Computer "brains" guide robotic "bodies."

In our era, the construction of mechanical people has intervened in public life; in defense policy, in industrial policy, in research institutes and schools, the robot is unexpectedly our new companion. The computer—a twin issue to the robot—even became a social and political issue during the 1980s. Schools, families, writing, workplaces, play, and therapy are only a few of the sites where emergence of computers occasioned a renewed dialogue about technology. When this early rage over computers subsided, a residue or trace remained—an interest in the technologies and images of humans absorbed into a machine world. As a policy concern (when automation displaced workers) and as a literary device (when cute *Star Wars* robots received new defensive assignments, in the Strategic Defense Initiative), robots joined computers in our political discussions. The robot, so to speak, stepped forward.

In this chapter, I am trying to relate this strangest of human-created mirrors to nuclearism, our most frightening creation. Literary criticism directs our attention to robots when it suggests that we pay attention to mythic dimensions, noting especially the exclusions, displacements, and absences that pervade any mythic discourse. Still, there is a resistance to taking the robot seriously. We try to avoid an unavoidable conclusion; the relationship of people and machines will now be permanently controversial and intriguing. Still, long after that development has become evident, the presence of the cyborg somehow embarrasses us. On the one (biotic) hand, the robot is too obvious and childish; on the other, it is too metaphorical and literary.

I need to be explicit about why the robot belongs in a discussion of nuclearism. In a limited sense, the robot arrives as a direct consequence of the disappearing warrior, discussed in the previous chapter. There is more to the relationship of technology and politics than the pace of action and decision that is forced on us by communications technology. If this were only a question of pace, it would make more sense to cry "slow down!" But today such a call would be an irony, a joke; we're far beyond the stage where our resistance to technology makes sense to us, even though the tension between development and autonomy has long been debated, as Langdon Winner has shown.[2] Instead, society's relationship with technology has changed so significantly that many theorists now raise philosophical issues—questions of agency, autonomy, intelligence, bodies, and, hence, the opacity or clarity of basic political choices.

In short, the robot is one of the vehicles we use to discuss the social and cultural transformations of our time. At the basis of such discussions is the issue of agency; if the bomb can end (or mutate) society, in what sense do humans retain responsibility and capacity for action? Donna Haraway's treatment of the cyborg (the central image of what she calls "my ironic faith, my blasphemy") confirms such an approach.[3] Even the name cyborg (for *cyb*ernetic *org*anism) always suggests an exemplary connection of human and machine. In the world of work and in the military realm of $C^3I$ (command-control-communication intelligence), the cyborg is the image that registers. It is the fiction that nonetheless maps "our social and bodily reality." Having identified the closeness, the intimacy of machine and person, Haraway can locate a crucial tension. "The relation between organism and machine has been a border war. The stakes in the border war have been the territories of production, reproduction, and imagination."[4]

Such "border wars" are of special interest now; the border metaphor relies on mapping and emphasizes disjunction. One such problematic unity is nature itself, as Haraway shows. "The certainty of what counts as nature—a source of insight and a promise of innocence—is undermined, probably fatally." The difference between natural and artificial, mind and body, becomes thoroughly ambiguous. "Our machines are disturbingly lively, and we ourselves frighteningly inert." Thus, in Haraway's argument, it is the cyborg—not ideology or structure—that augurs the escape from unities and origins, "a creature in a post-gender world . . . resolutely committed to partiality, irony, intimacy, and perversity."[5]

I offer the shifts and reversals rehearsed in this chapter as an odd, off-balance reminder. A political position we would speak like this would be a politics of the code, displacing all earlier politics based on nature, essence, or presence. The appropriate mode for studying this new, perhaps distressing, political form would be more literary than scientific. In this search for a code, what we find may be ironic, rhetorical, and far more tentative than any scientist could honestly be.

We could even propose that the bomb now has a body, seemingly more material than the mushroom cloud and more visible and accessible than the specific weapons that are only part of a larger "system" of nuclear warfare. Even if "merely" in code, *the robot is the body of the bomb*. The two technological phenomena that have most fascinated us in the last four decades are now explicit twins. In the now out-dated metaphor of rationalism, the computer is the brains of this operation, the bomb the muscle. In its physicality, the robot is the encoded sign of nuclearism.

This is an ambivalent development; robots are highly metaphorical and charged with mystique, but they do seem to represent some sort of worldly presence. The important political question contained in this ambivalence is whether this is an embodiment that might produce an actual critique—replacing the unfocused anxieties and false reassurances of other oppositions to nuclearism. My project, in this chapter, is to study the robot (up close and digital), to see how it might intervene into the landscape of disconnected signs and meaning I have begun mapping in my discussions of nuclear criticism and the warrior mythos. My claim is that these discussions of computer, robot, bomb, and self can best be understood as linked metonymies, mapping reversals, displacements, and desires too dark to confront and too powerful to discard.

## The Absent Army

In the previous chapter, I argued that the warrior has become an impossibility; now, we can examine a second layer of disappearance. The U.S. military in particular and "the nuclear threat" in general are characteristically opaque, somehow encoding an absence that nonetheless dominates us. We know that there is an enormous commitment by this country to funding and encouraging militarism. But military artifacts are remote, often incomprehensibly complex, and philosophically threatening. We have a strange militarism: few parades of troops and none of weapons, secrecy imposed even on devices completely familiar to potential opponents, and an economic analysis of military bases that focuses on retail activity in their neighborhood.

Opponents of contemporary militarism have often commented on this opacity, diagnosing an avoidance strategy. Nuclear weapons have such awesome potential and such an odd, inverse mission that we invoke a taboo, as Jonathan Schell explains. When we do see and touch mock-ups of the bomb, our reactions seem to reveal that this is a sacred artifact, not merely a patriotic one. In Los Alamos, New Mexico, the museum of the famous atomic lab displays models of the Hiroshima and Nagasaki bombs ("Thin Man" and "Fat Man"). The museum mounts these extraordinary artifacts—models, of course—in a thoroughly unceremonious manner, in the back corner of a bright and prosperous "visitors' center" that features historical information about the lab and educational demonstrations about nuclear physics. The gait and posture of visitors change when they near the models, even though nothing about the museum's design should evoke such a response. No children pose for snapshots on the nukes, as they might when visiting Civil War cannon or modern fighter plane displays at other museums. Visitors hesitantly touch the Los Alamos models as if they were icons, which they are; taboo icons, at that.

We thus have at least two predominant ways of dealing with images of militarism. On one hand, the most abstract symbols are held in an almost hysterically high esteem; there is the flag-burning controversy, of course, but also the pervasiveness of pseudomilitary fashion in almost every area of design. Stylists make flight jackets and pants, trucks, eyeglasses, and untold other products that seem to be military artifacts but are not. On the other hand, it is seldom transgressive, now, to avoid the symbols of militarism. The encounter with the Los Alamos museum suggests that we are made uneasy by the honest, straightforward display of what ought to be, by any prenuclear logic, patriotic signs. In other words, deterrence now does not even have simple, direct, and positive symbols, as the Richland High School students mentioned in my introduction discovered. Armies have long thrived on flags, uniforms, badges, and the other symbols of their presence. There has been some tension between the glory of these signs and the gore they represent, but this tension is now unmanageable. The Strategic Air Command's slogan, "Peace Is Our Profession," now strikes more than just pacifists as absurd.

This opacity (of both the military and the nuke) is explicit, reversing an old era in which the visible, massive presence of military forces reassured citizens by warning potential opponents. Now, the bomb threatens to disappear entirely. We can admit the hazy "visibility" of its effect—the mushroom cloud or the shock waves of the underground test. But the bomb is hidden from view, almost as thoroughly invisible as the radiation it would unleash. In contrast to its many predecessors, this weapon is kept away.[6] The military routinely "refuses to confirm or deny" the presence of such weapons in any particular location. The Soviet practice of parading their nukes during the May Day parade seems quaint and not quite civilized, even though parades of the military were an American staple for generations. (It is not surprising that Gorbachev chose the 1990 May Day parade as an opportunity to stage a demonstration. He is far too postmodern a character to suffer that display without some embarrassment.) Occasionally, antinuclear activists in the United States have surreptitiously photographed the movement of weapons on military bases. In the 1970s, such pictures became major front-page news in Honolulu. In an example mentioned earlier, visibility was an explicit issue in demonstrations against the Pantex weapon-carrying "white trains."[7]

The military's role in computer and robotics development has been as secretive as its other endeavors. These devices have been—at every stage of development—artifacts of militarism, first and foremost. Commercial use—the "do not bend, fold, spindle, or mutilate" billings by which computers first became publicly visible—spread computers throughout society, but obscured their actual development path. Popular histories of technology frequently focus on devices associated with consumer products such as the telephone, rather than on military electronics and air traffic control, both of which played a larger role in the development of computers. Bell Lab's work on the transistor is more widely known than that of Texas Instruments; we thought we knew the products Bell produced, and we conveniently forgot its role in the Safeguard ABM system that preceded SDI. Even though IBM (International *Business* Machines) is a major defense contractor and sold most of its early computers to the military, we perceive it as involved in commerce and education. Honeywell became controversial for something called the "automated air war" in Vietnam, not for the computers that made such a strategy possible. And Motorola makes TV sets.

By now, this ritualized opacity is so well in place that certain kinds of visibility can be incorporated without endangering the overall ritual. Or, perhaps, the computer and robot are now so pervasive that they must be both opaque and visible, at the same time. In either case, however, the robot and computer are recast as calculators, number-crunchers that could hardly be a threat. But even that reconciliation backfires, entering the map of social metaphor in an odd, discomforting way.

Ellen Willis gave expression to the "calculating machine" metaphor and linked it with several types of catastrophe, finding the common thread. Writing shortly after the spring 1979 implosion at Three Mile Island, Willis reflected on her unwillingness to calculate continually which new disaster scenarios seemed most likely and, thus, should motivate serious preparation. "In short, I lack[ed] the entrepreneurial attitude toward apocalypse."[8] Willis reports that neither she nor her friends took TMI seriously at first, and certainly they did not evacuate their New York city homes. This turned out to have been the right "choice," even though it was not really a choice, since the criteria were so ambiguous. Choosing right was no confirmation, no reason to be smug. The calculating, entrepreneurial attitude Willis wishes to evade continues to be called forth by political events. We have often been in situations that demand that great decisions be made before we have quite captured the categories of choice. Events of a certain type seem to place us in a continual pattern, trying to figure out how to respond. The only well-established alternative, mass political change, is undeniably "a long-range affair," offering little or no consolation when events such as TMI suddenly overtake our calculations.

Willis conceded that her response to TMI was "a kind of feudal fatalism. I identify with the people who lived on the slopes of Vesuvius." Amidst what conservatives see as America's triumph over the volcano, even activists sometimes seem comfortable with academic exercises of critique and entrepreneurial games of cynical calculation. We may forget that the earth will soon shake, perhaps today, and our act is still pitifully out of order. Are the volcano's rumblings real this time? These constant "rumblings" unsettle even the most politically committed among us. Even though the events of TMI turned out to be relatively minor, the episode reminded Willis how high the stakes might be. Remember those German Jews who stayed too long? She can identify with their judgment "that the Nazis' anti-Jewish campaign would be limited and temporary, that they could weather it." To leave what had become a comfortable, middle-class setting "must have seemed an unnaturally drastic response."[9]

This is melodramatic, and that's the point. The politics of nuclear technology is still unfamiliar, almost mythically hard to grasp. Such an issue — closely related to Baudrillard's catastrophe — invites melodramas of calculation (now with CNN's "Round the Clock Coverage"). It tests wits. We may choose wrongly, or our non-choices may be disastrous. At the same time, meaning, cause, and effect all have a notably remote quality; we know these speculations will be abstract—will remain unconfirmed, socially or objectively—until events provide their own verification. Rather than relieving us of the responsibility of choice, however, that quality forces us to shore up our doubts by lodging the choice in as "responsible" a discourse as possible. And calculation is, as we know very well, responsible.

Not only nukes, but also each geopolitical crisis that surrounds them (from Cuba in the 1960s to Iraq in 1990) place us in a continual pattern, trying to "figure out"—to calculate—how to respond. We sit before the tube, clicking the remote control as if it were the pocket calculator it resembles, figuring the odds associated with decades of racial conflict, Vietnam, recurring oil crises, and, of course, the Gorbachev era. Even if this pattern of waiting and calculating had characterized politics before our age, the contemporary era finds us reacting to dramatically new political spectacles, further from our experience but more vivid than ever. This is why TMI touched such a nerve; here was a genuinely new technology—itself abstract, experienced indirectly even by insiders—that had to be folded into our existing frameworks.

If the calculator metaphor was not quite enough, however, this late stage of modernity could come up with other possible dodges for the mess of ironies and perversities that overrun the computer/robot/nuclear age. In the next section, I raise the possibility of a social-learning, education metaphor, based on the thorough infusion of the computer into society. In the grand tradition of Rousseau, backed into the corner of a diagnosis at odds with any conceivable alternative, we could play education trumps.

## Techno-populism

The recent discussion of computer technology may not have seemed, at first, to be a coded version of the nearly half century of debate over the bomb. This time, the debate had a notably popular tone, not the dark neurosis of the coded repressions we have learned to expect of the nuclear age. The intrigue with computers was neither abstract, academic, nor entirely confined to elites. State legislators made funding for classroom computers a priority. Teachers at every level of education at least briefly shared a preoccupation that could only compare in its scope and urgency with the Sputnik fervor of the 1950s. Writers explained to colleagues that they had, finally, purchased a machine and could now hardly do without it. Citizens who might have been merely annoyed by pinball machines or jukeboxes at the local soda shop mobilized to force zoning hearings aimed against the video game arcade at the shopping mall. Having finally decided that a stop at McDonald's was a relatively innocuous bribe in the ambiguous struggle of contemporary families, parents found themselves deciding whether to give the kid a quarter with which to shoot aliens.

In short, a fascination arose as we watched routines of work and leisure transform at a pace so quickened as to seem palpable. Pop-tech "computer literacy" courses and simple familiarity (accumulating at a startling pace, with sales of millions of computers) may have softened our concern, but the political framework for our concern with computers was not obvious. While I understand that there are many ways to investigate the role of robotics — in terms of workplace politics, political economy, weapons technology, and so on — I have chosen to emphasize the cultural and literary (the textual, in several formats). This approach is not complete or definitive, but it does have the virtue of forcing our attention to the odd, displaced qualities of an issue with strangeness at its core. In American political culture, what is familiar gets defined as populist, but the invention of computer populism was *odd*.

Proponents of the microcomputer (or, as IBM dubbed it, the "personal" computer) argue that these things are distinct from their larger relatives. Computers became a public presence at the moment when processing capability became so inexpensive that home computers were possible. This was a public movement, not an elite, corporate one. The major computer companies and academic experts had consistently predicted home computing based on time-sharing with large, central computers, even after microprocessors (on which the home computer is based) had become available.[10] Acting on populist, hobbyist impulses, individuals marginal to the world of computer commerce created a "microcomputer revolution." It should not be surprising that the political populism of the earliest microcomputerists diminished as these machines became consumer items. After all, an ideology of "critical distance" or perspective requires new (or otherwise distanced) points of departure. Depoliticized and familiarized, computers soon lost whatever critical edge they enjoyed within such an ideology. The machines could be derrogated as just more commercial technology, either as commodity or productive technology. The corporate MIS ("Management Information Specialist") displaced the anarchic amateur experts and the machine — never mind how extraordinary — was reabsorbed into dominant ideologies.

In some ways, the computer may have conspired, all along, with this domestication. In retrospect, we might well conclude that whatever "microcomputer populism" arose never did have much ideological content, the experiments in socializing of information conducted by Lee Felsenstein and Berkeley's Community Memory notwithstanding.[11] Insofar as computers were ever populist, theirs was a populism of style — extra-corporate and at least intermittently cooperative. Such a populism could be "authentic," but it would inevitably risk a characteristic lack of durability. The belated but ultimately triumphant efforts of IBM, combined with the unexpected omnipresence of (frequently mundane) games, confirmed that the dominant ideologies of the computer also applied to microcomputers.

To seek a closer view of the computer, rather than a broader perspective where ideological critique might still seem possible, we would need to pursue the people who desired the machines most intensely, displaying in that relationship both affiliation and more perverse possibilities. In this case, the desirer is called (could Hitchcock be far behind?) "the hacker," suggesting a destructive, transgressive impulse. At the level of creed, these programmers articulated a privileged status for learning, giving discovery itself a pure preference over any competing interest, notably including property or power. The "hacker ethic" centers on an openness of learning and teaching, finding practical ways to elevate those practices over other (usually more powerful) social goals. This much of the description may make the hacker seem a creature of the Enlightenment and critical distance; in tension with that possibility is the rigorously antihierarchical, disruptive, even subversive tendency of the hacker and his or her hack. Unlike the Enlightenment humanist, the hacker could never hold "legitimate" power.

The hacker habitually relegates to lower status ownership, security, and the managerial impulse for hierarchy, security, and control. Their aesthetic flair privileges learning against the strong, centralizing impulses implicit in the technology they love.[12] What the many chronicles of the hacker often miss, however, is the congruence between this creed and the object of the hackers' desire: computers characteristically acquire codes and execute them, with remarkably little concern for the intent or context of the code. Both the successful program and the failure ("GIGO — garbage in, garbage out") characterize the computer. The hacker is an overdetermined sign, attached to the most dominant of technologies, but still retaining traces of subversion.

For a time, before the ideological response found a second wind, the techno-populism of the hacker threatened to go mainstream with this undistanced populism of desire, information, and indifference to consequence. Consider the instance of LOGO programming language inventor Seymour Papert, who issued the most aggressive statement of this populist impulse.[13] Papert even argued that the pedagogy enabled by LOGO actually implied the demise of schools. Of course, the rapid spread of LOGO produced no radical assault on schooling. Educators may have simply ignored Papert's presumption of computing's radical potential, responding instead to disciplinary practices more central to the schooling project. Why think of revolution when one could introduce "teaching materials" that would keep students attentive (and in their seats), while convincing parents that something pertaining to future employment happens at school?

Still, it is no shock to the hacker that things are ephemeral. After all, hackers deal with the most ephemeral events; information moves at the speed of light inside those black microchips. Even Papert's brief effort at political definition stands as an unusual and notable episode. The populism of early microcomputerists more often slipped into the utopian apologies of AI partisans such as Marvin Minsky, and pop-tech assurances of enthusiasts who insisted that computers were nothing to fear.[14] None of these apologies completely dissipates the characteristic edginess of the hacker, the asocial intensity of the nerd. The residue of unaccounted-for myth and desire is edgy and giddy, not assured and confident. Every reassurance is dogged by what it must reassure against—namely, an ambivalence that is not spurned. We remain uncertain about the meaning of these machines, long after the ideological critique has lost its best critical edge.

Looking at the computer from the odd angle of the hacker, issues of agency and autonomy displace Enlightenment concerns about responsibility and humanist hegemony. These displacements will relate to our investigation of the nuke, but before returning to that topic, there is another aspect of computers to consider. Theorists have recently emphasized the visibility of things, asking how something becomes visible or invisible, how an event might enter or resist the realm of the image (a field of representation). In this area, too, there is useful work to be done on the computer.

## The "Black Box" That Walks

Although technology has always promised power, this new technological populism goes further, fulfilling its own populist rhetoric by producing *expectations of openness*. Concern about the social, political, and pedagogical aspects of computers displays a special characteristic; although these machines are omnipresent, they are also opaque. Although the meaning of other machines and technologies may seem transparent, computers elude our attempts to categorize, even though they also exhibit the transparency of guileless, speed-of-light transformation. This technology is somehow obviously, almost "self-evidently," significant. We seldom doubt that computers will "change our lives," as the saying goes, but the technology is still opaque and inaccessible. We catch ourselves taking note, like the frightened child, that there really is no person inside the vending machine that talks.

Hackers use the term "black box" to refer to a self-enclosed mystery—a competent but seemingly opaque piece of machinery, encased not only in metal or plastic, but also in a puzzle. Taking their joke seriously, we find that the black box is metaphorical in an odd way. More precisely, it is transgressive in that it defies metaphor, pointing instead toward puzzlement and—crucially—absence (of explanation, meaning, or simple direction). Computers—the ultimate black boxes, since they can be transformed, within that case—register an opacity that diverts our attempts to locate their meaning, their standing in the world. The computer is not simply a subservient machine, nor is it (we hope) quasi-autonomous and, hence, pseudohuman.

Sherry Turkle investigated this opacity by observing the difficulty young children have when asked to categorize computers and electronic toys. Turkle observes children in the act of discovering problematics about self and knowledge, prompted by encounters with a computer. These things might be alive; after all, they have the ability to surprise. They may (or may not) be self-aware, which may (or may not) be important.[15] Turkle's subjects are on to something; these things are not mechanical. They refuse to offer the reassurance that if we disassembled them we could eventually understand how the gears and pulleys work. On the contrary, computers have a maddening lack of moving parts. Only after a leap of imagination (of philosophical categories) could we begin to call these things mere machines—explicit parallels of our automobiles, sewing machines, and typewriters.

As I suggested earlier, this opacity is all the more troublesome since computer technology (or, as it is called, information technology) promises clarity. Lacking machine noise, emerging from industrial sites (always located in oxymoronic "industrial parks") that are absurdly lacking in the dust and grime we just know must characterize a factory, the computer age is so light and clear that it edges toward invisibility. Haraway emphasizes the clarity of the disappearance: "Our best machines are made of sunshine; they are all light and clean because they are nothing but signals, electromagnetic waves, a section of a spectrum. And these machines are eminently portable, mobile—a matter of immense human pain in Detroit and Singapore." Whether the machines are opaquely absent or clear as they disappear, the implication is the same. As Haraway notes, "The ubiquity and invisibility of cyborgs is precisely why these sunshine-belt machines are so deadly. They are as hard to see politically as materially. They are about consciousness—or its simulation."[16]

There is more to this hiddenness than a lightness or a lack of moving parts, even though these attributes are important. The opacity of computers also provokes reconsiderations of scale and transformation. The mobility mentioned by Haraway is a function of a tendency toward miniaturization that also contributes to opacity. *The Wizard of Oz* would not have been such a reassuring allegory if the drawn curtain had revealed a microscopic Wiz. The watch-on-a-chip and tiny pacemaker are at least part computer, and the notebook-sized "electronic workslate" portable computers are, unquestionably, fully computers. At the other end of the scale, the computerized Cheyenne Mountain facility of the Strategic Air

Command, looming over Colorado Springs, is described as an integrated, complete, national airspace monitoring system of huge scale in both size and effect. If it is the "CPU" (central processing unit), most of the continent is its "peripheral." This diversity of scale is unique in our accustomed world of things. A tiny model of a car—or a huge filmed image of the same car—differs in scale from the actual car, but the model cannot be confused with its object. There is no such home image for the computer; boundaries shift and the notion of a core object becomes harder to focus.

This flexibility of scale contributes a layer of opacity; what could these outrageously diverse things have in common? When we investigate that opacity, however, we soon find yet another layer of opacity at "the heart of the matter." The computer's core not only lacks mechanical parts; we cannot even see the working parts of a microprocessor without extraordinary (and confusing) magnification. Haraway's take on this miniaturization emphasizes the ubiquity it makes possible. Microelectronics are essential to virtually every modern machine. Accordingly, these machines "are everywhere and they are invisible. . . . Modern machinery is an irreverent upstart god, mocking the Father's ubiquity and spirituality." Our personal experience of the machine changes, reversing the terms of modesty and giantism; the microprocessor creates the possibility of the tiny-but-deadly. "Miniaturization has turned out to be about power; small is not so much beautiful as pre-eminently dangerous."[17]

The scope of computer functions also assists this odd invisibility, hindering our attempts to focus. The computer is the most flexible of technologies, in terms of the variety of operations performed. And this flexibility is an essential characteristic of computers, as their proponents constantly remind us. Again, children have modeled a fascination with flexibility. Perhaps the least understood recent toy fad was the (well-named) Transformer, a small model robot whose only function is to be manipulated into a physical form that hides its robotic appearance. In one example, a robot transforms into a model of a heavily armed space ship. Although some parents were surely relieved that Jane and Johnny no longer wanted simulated Uzi machine guns for Christmas, the Transformer taught lessons, too. The robot is mythically flexible, capable of unimagined mutations if only one accepts that those changes are finite—planned in advance by some unknown designer.

Incorporating this opaque, shifting phenomenon into our existing categories is clearly not a trivial task. This technology makes us uncomfortable, in large part, simply because it confounds categories, eluding our grasp. Going even further, we begin to suspect that these machines might think. Despite a long-standing record of overstated goals and underachievement by Artificial Intelligence (AI) scientists, we seem compelled to take AI very seriously, even when it becomes clear that the possibility of pseudothought informs only a few of the policy questions involving computers. I cannot venture far into that argument here, but doubts regarding the status of machine "thought" recur often in our efforts to sift through these diversions and opacities. Skeptics such as Dreyfus and Searle have aggressively criticized the philosophy of mind implied (perhaps, required) by AI.[18]

More questions are raised by this portrait of a technology so opaque, flexible, shifting, and elusive. These are characteristics we would usually associate with a lack of agency and responsibility. In a sense, these are not surprising characteristics; they are, instead, precisely the elements we would expect to be raised in a managerial age, when jurisdiction battles erupt on the scene of any disruptive outbursts. We should expect these issues of control, and we would probably predict that such a recalcitrant (read "dangerous") technology be heavily controlled. It is toward that central political concern with control that I will turn next.

## Robot Power

The issue of control is central to our talk of robots. "Power" and "control" are pervasive terms in the discourse surrounding computers, appearing repeatedly in almost every article, advertisement, and instruction manual. As if to reassure, designers equip computer keyboards with "control" keys, as well as an "escape" key. Tropes of capacity and speed are routinely conflated with power—"this model is really powerful: 200 megs of storage, four megs of RAM, and a '486 chip." Proponents of the new technology routinely promise untold opportunities to automate the annoyances of work and life. Opponents warn of easily implemented, almost casual surveillance at a level previously unimagined.

Proponents of computers emphasize the role of human choice in the controlling of any "tool," and thus easily fit the relationship between "person" and "tool" into a discourse of addiction or pathology. It is our choice and we'd better choose well, else the techno-bogeyman will get us. But this will happen only if we choose poorly or abdicate our responsibility to impose a discussion of values on our day-to-day activities. Critics portray the fascination of hackers with solipsistic, neurotic control, abused by this technological engagement. They see hackers as damaged goods, people unable to function in social contexts, foreshadowing what would happen to us all if we fail to preserve human values. Theirs is a traditional refrain, heard even when Byron addressed Parliament to question the level of force used against Luddites, who he thought might foment an overwhelming social revolution by breaking those looms that would surely turn users' bodies into unskilled machine appendages.[19]

When we give computers either fictional or mechanical bodies, we reveal a fascination not only with technology but also with its twin topic, control—more specifically, the internal control of bodies by distinctively contemporary "selves." Our choices do more than reveal us. They reflect an already given discourse about control, defining a range of possibility. In literature, robots frequently undergo a familiar pattern of control. Frankenstein's monster became aware of his independence—he could move autonomously—experimented with it, offended the community, and was pursued so that he could be subjected to (bodily) control. In an ironic twist, the monster even tried to negotiate terms for a controlled, disciplined range of activity, a negotiation his creator could not face.[20] The predicament is a singularly modern one. Heroic science joins human responsibility in a pair that then begins to run amok. An oppression dilemma is enacted, and the participants know they have helped constitute the epic scene, to a tragic end.

Our hope is for machines that could offer unthreatening help, but the hope is shadowed by the fear that these new subjects might rebel, somehow renewing issues of domination, of politics. Perhaps concerned that machines could displace us with their "mental" as well as physical abilities, we anticipate help of the kind bodies can lend. We simply presume that technology should be under firm human control, and thus settle all technological issues by metonymically assuming that technology *is* under such human control. While Mary Wollstonecraft Shelley could focus on the doctor's torment and guilt, modern writers have had to confront the issue of technological autonomy, which is also the key to the contemporary popularity of revised Frankenstein fables in which the torment of the monster is finally sad, and the doctor is fully monstrous. Executing this switch, modern retellers of this fable establish the grounds for our own epic tragedy. We fear the subservient machine might become autonomous, but we have also witnessed the surpassing of Dr. Frankenstein in the technology of surveillance, control, and discipline.

In this century, as Michel Foucault argued, the control of bodies is a central fascination. Without agency or conspiracy, power operates on bodies, grafting its concerns onto them. With apprehension and respect, we grant the computer a special status when we give it a body. Those bold enough to postulate the autonomy of machines—Jacques Ellul on one side of the polemic, proponents of AI on the other—soon find themselves under critical attack. When the editors of *Time* made the computer "Man of the Year," they both participated in this popular embodiment and suffered criticism for that hint of autonomy. We will insist that our computers have robotic bodies, thank you, but the easier the availability of a power switch, the better.

This nervousness about autonomy suggests that the control at stake in the era of the robot is a different kind of control, unfamiliar to earlier theories. The robot controls itself, after a programmer installs the appropriate rules. It is this aspect of robotic discipline that makes these things the archetypical mirror of the postmodern self. Robots displace the factory, which had served as the modern age's mark of control. The steam-driven factory machines of the nineteenth century represented an externality of control. The most important political relation engaged by the self was with an externalized managerial apparatus of supervisors, bourgeois managers, capitalists, and, eventually, assembly lines and Taylorite efficiency experts. As Marx showed so convincingly, these features of social life were alienating; they placed into external centers authority that had previously belonged to the self.

Although Cartesians wished to conceptualize living things as reaction mechanisms, and despite the role of Pascal and Leibnitz in the invention of calculating machines, we are on entirely different ground now. There is nothing natural about the robot, nothing inevitable about the path this fable has taken in development. Clearly, we could have formulated other strategies for coping with opaque, mystifying computer technology. In another age, this ensemble of techniques could have emerged differently. The Age of Enlightenment, for example, might not have needed to embody the computer at all, had that era witnessed the development of mechanized calculation. We now celebrate the nineteenth-century Difference and Analytical Engines of Charles Babbage as precursors to the computer, and Babbage's machines reflected the factory-machine images of the Industrial Revolution. Other metaphors for machine thought were quickly subsumed, however, when the computer-as-person possibility emerged in this century.

The shift of control metaphors that accompanies the emergence of computers then becomes more than just another representation. The displacement of people by machines raises a dizzying, discomforting possibility, articulated by Jean Baudrillard. Robots are "the ultimate and ideal phantasm of a total productivist system," the expression of the goal of automation that is, in turn, what made both capitalism and Marxism possible in the first place. But that goal symbolizes a "self-devouring" system, in Baudrillard's phrase, since the technology making that system possible cannot be present at its ultimate fulfillment; people "are necessary for any system of social control and domination."[21] But the system of domination also contains its own self-devouring element, in the replacement of workers.

The economic realm critical theorists hoped would still provide totality dissolves when the robot obtains the status of master metaphor *and* defender. The robot completes the detachment from any "real" economics, the demise of which Baudrillard had already foreseen. Simulated arms and legs do the "labor." Fuse-changers and oil-can carriers absorb the displaced *labor value* (once engineers complete the robot's design), but they too are quickly and inevitably automated. The machine-like boredom and terror of work is finally just given back to the machines. Disorganized efforts to blame Asians for our unemployment only disguise the underlying terror generated by these shifts. This time, unemployment may actually be terminal, not cyclical.

In a typical Baudrillardian twist, politics itself is what finally disappears when work has gone. All fables of objective interests at war with one another slip away, replaced by (a finally pointless) technological management. Since the beginning of the liberal era, the goal of a productive efficiency has been the code for progress in general. But if automation means the displacement of people still needed to maintain the symbolic drive to efficiency, Baudrillard's possibility—a code for disappearance—exists. Industrial (or literary) production could now simulate a process of humanity devouring itself, a far more perverse possibility than any Marx ever entertained. As strange as it is, this outcome mirrors the nuclear outcome and the Frankenstein plot. At the end, people observe that their best and brightest products had been canceling futures all along. In such a context, the resistance to the "nihilism" of Baudrillard, Foucault, and Nietzsche could be simple self-deception, another simulated, intimate negation. As I will suggest in a later chapter, the further push by Star Wars toward this odd outcome (in the name of promised safety) only completes the scene.

To be sure, the robot *is* sometimes portrayed as a simple, old fashioned oppressor—as alien—in the sense of an externalized foe of the interests of class or liberty. That portrayal betrays its own peculiarity, however; such roles are so atypical that fictionalized accounts almost always call such machines androids to mark the difference. Real robots act as mirrors or metaphors of paradoxical, calculating independence. The key to what makes a robot, once designers (who also can now be automated) create a body, is the device's internalized discipline. Science (and fiction) writer Isaac Asimov summarized this discipline early on, in the "Rules of Robotics," which—he takes pride in reporting—have become an enduring subtext of scientific robotics, as well as of science fiction. The rules are hierarchical and general, and they are *rules*, not instructions. The robot must calculate, and temper its responses accordingly.[22]

The 1987 film *Robocop* was structured entirely around the interplay of these rules, as implemented by a corporatist police/state onto the half-man, half-robot title character. Robocop—and his generation of fictional and nonfictional relatives—had to be independent, but also had to incorporate self-control. Any robot who goes awry does so, by definition, when its ability to monitor itself falters (or, in Robo's case, when he finds himself in a contextual double bind with respect to the rules). That is the source of authority that encompasses them, imposed by their programmers, but also imposed by our patterns of thought and speech on the topic of agency and technology. This byplay makes movies such as *Robocop* for a simple, if spectacular reason. This is now our (human) authority, too.

My reading implies a record of misunderstanding. The threat is not that robots would become so successful as to displace humans, but that humans would unproblematically simulate their robotic mirror. Put another way, if we could get all the robots to behave well, we would not be stunned to discover that they were still robots. Absorbed by nervous defensiveness, however, we might not notice how much like them we had already become. We also should remember, however, that the special status of robots does not reflect a simple, empirical functionalism. Hopes and fears about robots and computers continually race ahead of what these things actually do. That is to say, this status is characteristically fictional, or mythic.

Critic Hubert Dreyfus is at his most persuasive when he argues that computers cannot be embodied and hence are incapable of intelligence in any meaningful sense. But Dreyfus's judgment has been an unpopular one, hardly phasing even those who best know this technology. As Dreyfus notes, AI scientists have continually acted in a manner unlike more cautious and more traditional scientists. These new scientists have always been promoters, too. They have routinely made impossible promises, and have exaggerated their successes; time lines become almost entirely fictional. Even the title of their discipline, "artificial intelligence," is ambitious and rhetorical, and their experiments have carried such grandiose names as the "General Problem Solver."[23] AI science seems only recently to have discovered the tentativeness that usually characterizes scientific talk. It is as if the opacity of computers has influenced even those people most familiar with them. As was also the case with the alchemists (whom Dreyfus calls the precursors of the AI scientists), the opacity of the objects of their research reminds these scientists that everything is possible.

There is another way to understand these failures, and this way leads toward the linkage of robots, computers, and bombs that I am trying to evoke. One suspects that the failures have been strategically useful, undercutting perceived danger and reassuring us of some intrinsic human worth. Luckily, as we say, these things never seem to work right. This strange reversal of the cult of efficiency and functionality should be a clue. Trying to use this clue, we may begin to suspect that the robot is a kind of code, a symbolized reality in which function is somehow reversed. In *Robocop*, one corporate faction treats a spectacular robotic failure as hardly more than routine; if the failure contains enough spectacle to engage contractor interest, only more orders for upgrades and spare parts will follow.

Even while *Robocop* was still showing in the nation's theaters, the "news" confirmed the fiction. A nationally broadcast radio program featured an interview with an anti-Pentagon lawyer who reported on the attitude of defense contractors; "The attitude is, you won't ever have to use these things. And if you do, it doesn't matter."[24] This attitude offends the activist lawyer; things should *work*. But the attitude he reports is an implication of nuclear deterrence. Others have been clearer on the issue of functionality, but their tone is almost always exasperated. Seymour Hersh reports that the issue first arose when the United States planned to install Pershing IIs throughout Europe in the early 1980s. Reports that the missiles might not work did not matter. An Assistant Secretary of State for European and Soviet Affairs told a staff meeting, "We don't care if the goddamn things work or not. . . . After all, that doesn't matter unless there's a war. What we care about is getting them in."[25] The Pershing was more important as a sign of European cooperation (or submission) to Reagan weapons deployments than as a sign of destruction.

As twisted as this reversal is, it also should be familiar, embedded as it is in the reverse logic of nuclear deterrence, where things routinely exist in order that they *not be used*. Amidst such an inverted logic, competence has new angles. Diane Rubenstein has noted how well "giddy incompetence" served French foreign interests in the Greenpeace affair.[26] Such reversals should suggest that we are gazing at a highly symbolized scene, perhaps even a simulation, to follow Baudrillard's suggestion. Perhaps robots are the embodied stand-ins for a power that absents itself, perhaps because scrutiny of it might be dangerous—to us, but also to its hold on us. Robots could be the simulations that provide as much focus as a simulated political life will need. And they might not be entirely humorless in this role, teasing us citizens of the "real world" now and then about our convictions that "things have functions," "exist to be used," and can be known and judged in that simple context of use.

## Desire, Absence, and More Control

The psychological interventions of Jacques Lacan have been appropriated by postmodernists to help raise issues of metonymy and desire as substitutes for the naturalistic concern for metaphors and symbols. Lacan's work is difficult and resistant to this sort of "use," but it seems possible that some of his concerns might help untangle these odd, coded opacities. The resistance exhibited by computers through opacity, the fascination with which we nonetheless react to them, and the "search for self" grafted onto them by AI scientists—all of these could be reinterpreted.[27] In Lacanian psychology, this fascination and lack raise questions of desire. This opacity signals more than "lack of clarity," or imprecision, but it does not necessarily signal metaphysics. Lacanians suggest another possibility; given the centrality of desire and the unlikelihood of its fulfillment by this opaque set of things, we are embroiled in the psychology of absence. Metonymy is the rhetorical trope of absence and desire, by which we substitute one word for another, more literally appropriate choice (or the word for a part stands for the whole). For Lacanians, the name change follows a trajectory; metonymy forms a procedure in which causes express effect, contents express container, or part expresses the whole.[28]

This is my claim, then, in response to every attempt to strain meaning from these opaque machines: *the computer and the robot are the metonymic processes we use to deal with the nuke*. If the nuke has seemed unspeakable, then, we may have been using a metonymy. The computer and the robot—parts, implements, and artifacts of nuclearism—compose a Lacanian trajectory of desire. Displacing questions of censorship (and transgressive responses to its prohibitions) with possibilities of mastery, assimilation, and accomplishment, we deal with Lacan's "desire for something else," with metonymy.[29] Programming the computer or robot, we have controlled the (uncontrollable) bomb. Familiar with our personal silicon, we simulate some kind of familiarity with remote isotopes controlled by forces inaccessible to us. We have walked, seemingly fearlessly, into a (silicon) valley of doom prohibited to us.

As Deleuze and Guattari emphasize, this displacing process continually breaks away from the ego, canceling it with several identifiable processes.

> The law tells us: You will not marry your mother, and you will not kill your father. And we docile subjects say to ourselves: so *that's* what I wanted! . . .
>     There we have a . . . displacement. For what really takes place is that the law prohibits something that is perfectly fictitious in the order of desire or of the "instincts," so as to persuade its subjects that they had the intention corresponding to this fiction.[30]

Deleuze and Guattari repeatedly use that exclamation—"So that's what I wanted!"—to bring a metonymy to the fore. This process can be applied to our discussion of nukes. Faced with the Law, the egoistic expression of values, we exclaim, "nobody could possibly want nuclear war." But, unmistakably, war preparations are everywhere. The reverse logic of deterrence—the weird mission of preparing nukes never to be used—comes home to roost. Desire breaks out of the censorship planned for it.

Even if nobody wants nuclear war, people are acting as if they do. So *that's* what I wanted! In this way, the simple repression of nuclearism that liberal humanism tries to turn into a first value, a base agreement, twists out of control. A metonymy forms under the sign of the exclamation heard constantly in American political talk over the last forty years: "Nuke 'em!"

The robot, then, may be the ultimate nuclear-age metonymy. We "desiring machines" (as Deleuze and Guattari refer to modern selves) produce a mirror image onto which these desires and displacements can be grafted. Desire is a feature of bodies, so we give the nuke a body. The opaque machine—a machine-without-parts—becomes a body (in Deleuze and Guattari's schema, a "body without organs"). Rather than making the leap to conclude that all the puzzlement of this new technology must imply a meaning at the core of human life, we could turn the equation around. We could reinterpret the projects surrounding electronic technology. And when we do that, we can hardly miss that a rage for embodiment has been accumulating. Computer partisans, critics, and popular artists have all used this embodiment as a way of dealing with technology; computers had to be embodied.

Sometimes we can glimpse this embodiment project right in the middle of the computer-ontology camp. When the MIT authors of the educational programming "language" LOGO sought to enhance learning by devising an easily accessible entry into a sophisticated computer language, they began with a robot. Students learn LOGO by manipulating a "turtle," which is either a mechanical robot that moves according to programmed orders, or an electronic representation of such a device moving on a video screen. Teachers encourage those students who have difficulty with this "turtle geometry" to plan turtle movements by moving their own bodies and reflecting on those movements. Watching LOGO students in their dance of angles and steps, we begin to see this conjunction of computers and robots.[31] Bodies have been appropriated and the opacity begins to clear.

In short, all robots—fictional, pedagogical, or industrial—enhance the meaning of otherwise opaque and elusive machines by bestowing on them an identifiable (if still amazingly flexible) embodiment. Educators confirm what horror-movie producers, editors of business magazines, and toy designers also know; robots are tremendously popular. Surely, this fascination with "mere tools" represents an attempt to reflect upon something that resists more direct consideration. Specifically, when we grant the mythically opaque computer a body, we are addressing political concerns. We know that bodies are subject to control.

In this context of body-and-control, the robot gives play—representation, even if in disjointed, odd images—to a breakout of desire. It is our very own "second death"; having set the robot up as our double, we could gaze at its demise. As Deleuze and Guattari emphasize, desire does move in that direction: "Desire desires death also."[32] Humans do *make* robots, but it is a hyperproduction. Embodying this technology, we have produced a projection screen onto which we can display our version of the production-desire saga. In a rush of postproductive enthusiasm, we finally surpassed Lacan's mirror stage of development, but only by externalizing our ability to project. Suddenly, it is the robot that desires the nuke; we have become victims of autonomous technology, without wanting to admit that this is possible. But the displacement is still functioning; gazing at the robot, we are still looking at (through) a mirror.

The consequences of this robotized death wish are multiple, and I will only review two, implying some range of operation. In the first, the displacements of desire map onto whole areas of public policy that otherwise seem far away from the nuke, connected only through the indirect but all encompassing linkage of fiscal constraints. The second poses consequences more personal in character, playing out Deleuze and Guattari's "So that's what I wanted!"

In the first case, Dean MacCannell has added to nuclear criticism's understanding of power in a context dominated by "attempts to bring an imagined future moment, a moment of awful destruction, to bear upon the present."[33] After reviewing familiar dominations, MacCannell identifies a possibility that would truly resolve this dilemma. "Taking Lacan's suggestion seriously and treating macrosocial arrangements as expressions of collective unconscious desires, beneath the surface of fear, the supposedly unthinkable prospect of millions of instant deaths . . . one can find evidence of a growing desire to experience the bomb."[34] At the margins of contemporary life, we can witness another side of the displacements and absences I have been evoking. Self-control, loss of will, and calibrated, calculating minds are only part of the postmodern setting. The other side of that self-control—an erotic side—is the urge to experience the bomb. It could be that, at some level, we have already started to develop a surprising response to Stanley Kubrick's injunction that we "stop worrying and learn to love the Bomb."[35]

MacCannell argues that one could interpret contemporary culture as thoroughly infused with an unprecedented death wish.[36] We are preparing for nuclear suicide. In his central example, the obviously negligent treatment of American cities finds a meaning—through some poorly comprehended political process, we are preparing them to be offered as sacrifice. His argument is a reminder of how broadly nukes act on us. By extrapolating our criticisms, consciously divorcing them from the discourses they intend to critique, we can start to identify the effects of this culture.[37]

This represents a rhetorical way to represent the implications of accumulated policy decisions of the nuclear age, including policies seemingly unconnected to nukes. MacCannell finds in this approach a way to articulate the "death wish" many nuclear opponents have cited in less explicit ways. "If we only go so far as to maintain the poor in the cities, we are in a no-lose situation. It is America of the Hollywood happy ending after all. If the missiles come, they will wipe out all our problems. If they do not come, [capitalism] can once again flourish."[38]

In the second case, displacement finds a path for desire in personal, psychological terms. In the 1986 film *Desert Bloom*, displaced desires of every sort (from incest to genocide) threaten to burst out of the 1950s Las Vegas–*Atomic Cafe* setting.[39] It would be trivial to say that these warring desires make the characters of this drama "unhappy," or that they would be better adjusted if they docilely accepted the nuke. More to the point is that these displaced desires collide and permeate lives that come to focus on the most apocalyptic violence at hand—the nuke tests beginning just outside of town. Disarmed, the characters displace catastrophe onto unknown and absurd possibilities somehow promised by the tests (and then "confirmed" by small town rumors, nuke drills at school, and the backdrop of the Korean War). When the blast happens, a radio announcer breaks the silence with the truth; "Well, the A-Bomb went off and we're all still here, folks."

Near the end of the film, the storyteller (thirteen in the film, but telling the story as a reminiscence from later in her life) tells us that her abusive, alcoholic, bigoted stepfather would do better after the awful events of 1951, but given what has transpired, we can only wonder what "doing better" would mean. Deleuze and Guattari offer a suggestion:

> Absorbed, diffuse, immanent death is the condition formed by the signifier in capitalism, the empty locus that is everywhere displaced in order to block the schizophrenic escapes and place restraints on the flights. The only modern myth is the myth of zombies—mortified schizos, good for work, brought back to reason.[40]

## The Darkest Code

The robot—this symbolic mirror that so attracts our gaze—has, in turn, helped us reconstitute ourselves. The calculating machine that moves and "manipulates objects" is no mere toy or diversion; it is more like self-description. Perhaps the first characteristic we recognize is this perpetual calculation, the commitment to metaphors of decision making that rely on constant, incremental comparison of binary possibilities. But we also can suspect, in the occasional gaps made possible by art or insight, that we have been flattened (without a blast) by the nuclear age, that we have long since become the zombies Deleuze and Guattari evoke.

A clear understanding that the dangers are "real" could yet mobilize responses. But, as critics have charged, this postmodern insight into the disappearance of "real" responses *is* so serious an occurrence that politics is debilitated. The criticism of Foucault—that he makes politics disappear—turns out to be right, but it wasn't Foucault's doing. Arthur Kroker and David Cook have collected several of the keys to this debilitation:

> We are deep into Nietzsche's cycle of the 'in vain,' the downside of the cycle of cynical power, a time of the most intense disenchantment with the psychology of sacrifice, that, in medieval times, went under the sign of *acedia*. *Acedia* meant a sudden loss of the will to go on, a mutiny of the living body against a cynical power. Baudrillard tells us we're deep in the cycle of *acedia* again and the certain sign of its presence is what Barthes described . . . as the contagion of "panic boredom" which spreads out everywhere. . . .
>
> Consequently, while Marx may have analysed capitalism in its bullish phase, Baudrillard's thought begins with the instant inversion of the 'circuit of capital' into the cycle of disintegration, exhaustion, decadence, and 'viciousness for fun.'[41]

Thus, while Marx "analysed capitalism in its bullish phase," Baudrillard began by inverting "the 'circuit of capital' into this cycle of disintegration." This amounts to "a desperate search for a revival of the real (real people, real values, real sex)." The "desperate fascination with the real" that marks the current period arises "because we live with the terrible knowledge that the real does not exist anymore, or, more precisely, that the real appears to us only as a vast and seductive simulation."[42]

Even if we join postmodernists in their despair of finding a political posture that will surpass such panicked, bored entrepreneurship, we can still go further in setting this context. Venturing away from that "desperate fascination with the real," we might yet begin to describe our situation in terms that fit, thus deriving some solace from this act of describing. The robot could emerge as an odd but compelling misrepresentation (a creative misreading) of our very contemporary selves. It is a misrepresentation only in the sense that is seems to promise the undeliverable, and even that sense is undermined when we notice the grid of desires in which the nuke is lodged. At its simplest level, the robot-as-representation only repeats a defense of humanity, in the classic terms of humanism. The liberal and humanist claim is that people are still responsible for what computers do; we have no "revolution," but rather another call for contemporary society to exert its values, to choose its future consciously, as it confronts yet another technology that has the capacity for either benefit or disaster. These "mere tools" have no more meaning than that we choose to give them. Agency is not at stake. The computer and the robot are not autonomous; they take orders.

This objection is more a reminder of our uncertainty about human control of technology than a reassurance. Having situated the desire to experience the bomb—as well as the routine adjustment of human lives to approximate the robotic—we can identify this technology as a different kind of tool. More than a reminder of Jefferson's dumbwaiter or Locke's blank slate, these machines evoke Baudrillard's mirror. The robot is our special mirror, a reflector we can send into the inferno, with its camera rolling until it melts, Dali-style.

In Huyssen's interpretation, Fritz Lang's *Metropolis* implies that the conflict of labor and capital could be solved by technology, even if the conflict proceeds on a highly metaphorical, mythical level of fear and hatred.[43] We could find similar connections with the nuke. The robot brings the horror together with the hope of controlling (always admittedly uncontrollable) horror. The postnuclear robot is a manufactured body; watching it, we are practicing the gaze that (as Foucault explained) now denotes control. True, the robot could overwhelm the observer. But that is the test we wish to arrange for ourselves.

The humanist insistence that we conquer this situation with expressions of will informed by value fails to comment sufficiently on the production of that choice. There are activists who understand the absurdity of Nancy Reagan's "just say no" in the context of drug use, but do not notice that they replicate that injunction when they say, in effect, just say no to nukes. Political response must be more than will; issues form important combinations, even if the linkages are not necessarily conspiratorial or structuralist. Having constructed this "choice," it will not do simply to choose.

At the margins, critiques such as Kroker's make possible a nostalgic stance of alienation. Models for unalienated responses must be created, and the computer-as-bomb presents such an opening. Rather than seeing each political question as an opening for humanist complaint, we could move on—light on our feet—to take apart another form of contemporary authority. Without a doubt, this critical activity will have political consequences.[44] The opponent's role is to continue criticism; each event can be posed as a fascination, not just a curt dismissal. Fascination keeps questions open, allowing criticism to continue. In the case of computers, we have widespread social experience with just such a displacement of alienation with engagement. From the position I have tried to evoke in this book, the privileged stance is proximity or engagement, rather than critical distance. In these terms, then, the uncertainty of our talk about computers has to do with our posture toward familiarity with objects that have been lodged in a discourse of renunciation or taboo. We have another "guilty pleasures" problematic here, one that is as political as it is psychological or literary.

When Birrell Walsh juxtaposed two metaphors for computing, he was exemplifying just such an approach:

> To those who don't like computers, these machines seem to be a kind of monkey trap—a bottle with fruit in it. The monkey reaches in and grasps the fruit, but with the fruit in his fist he cannot get his hand out. The hunter catches him because the monkey is unwilling to let go of the fruit. It is not our hand that we put into the computer, it is our attention. . . . These machines hold attention like no prior machine. To those of us who love these machines, they are a portal into . . . a puzzle-world, full of possibilities. . . . We agree we are absorbed.[45]

But, Walsh goes on to say, we could class this absorption as obsession and ecstasy, an altogether different context than traps and medicalized addiction. It becomes obvious that the real source of the critique levied against this engagement was an apprehension about an intense involvement with learning. My characterizations in this chapter form an attempt to shift the critical enterprise; the pervasiveness of the bomb can be acknowledged and related to actual human practice, and the appropriate response becomes critical engagement—even risk of obsession—mediated by "humor, vice, and the attempt to manifest vision."[46]

Rather than assuming that the spread of technology (and its multilayered opacities) portends meaning, we might find out more about both the machines and ourselves by emphasizing the puzzling opacity rather than the meaning. The leap to metaphysics is not necessary, and there are surely good reasons not to make it (as Derrida argues so convincingly). Rather than seeing metaphysics, we might let technology now remind us how complete the demise of any fundamental "essence" already is. Here is a phenomenon that *obviously* has no nature, except that it spreads, absorbing whatever it finds adjacent to it.

Instead of moving too quickly to denunciation, we could insist on continuing to practice opposition. If this insistence is based on a moral judgment, it still manifests itself in very distinctive intellectual approaches. The radically constituted dimensions of the situation could be (continually, radically, newly) demonstrated for "how they work," rather than for "what they are." That's deconstruction.

# 4

# Star Wars and the Freeze

The micro controls of the nuclear era are not the only form of power emerging at the end of modernism. Therapeutic and disciplinary discourses play a social role; they are, in effect, new agents that both constrain and produce contemporary political possibilities. These new practices of power interact with thoroughly constituted, nuclear-age selves. In the mix, there is a solution. The instability and perversity of the nuclear age no longer imply the fraudulence of humanity. The nuke is not a code of our failure; in a stunning reversal, it stands for progress. The chaotic paradoxes of our time carry the mark of a specific, contemporary control.

The instability of nuclear discourse thus works in two directions. As long as nukes remain unproblematic, both the repression of the deterrence paradox and the maintenance of micro controls continue to function. From a post-1989 perspective, we can begin to understand that we have been at an intermediate position, standing momentarily (if for a half century) between the pre-Hiroshima realm of military realism and modernism, on the one hand, and on the other an emerging postmodern realm where reversibilities, literary production, and the preeminence of pace (speed, communication, and image) fully and openly enter the political realm. Thus, while the nuclear age sorted out its politics, systems of power aggressively reinforced each other; antinuclearist claims validated nuclearism and vice versa. More important than the ebb and flow of what only seemed to be an antagonism, a new status and revised techniques of power just began operating and establishing themselves. This new power oppresses, to be sure, but it also acts positively and quietly, diminishing the possibility that politics would survive at all. As the age of deterrence gives way to the end of modernity, then, a struggle emerges, with the advanced methods of power encountering the new openings for politics.

There are, no doubt, many points of emergence at which this new struggle could be glimpsed. The example I address in this chapter is perhaps too obvious and overt, but we could still understand the politics of the freeze and Star Wars—widely regarded as excessively simplistic and obvious—in quite a different context, one in which they emerge as problematizing discourses, as characteristic of their age as is Reagan.

## Reaganism

When the Reagan presidency began, antinuclearists anticipated a different kind of trouble than they eventually encountered. The decade moved along somewhat unexpected lines; oddly, it contained overt hostilities yet ended in serious arms control negotiations. In an almost fantastic switch, Reagan enthusiastically joined the arms control project, even though he had long opposed such negotiations more steadfastly than any other man elected president or even nominated by a major party in the nuclear era. In the process, United States–Soviet dialogue was transformed along boundaries suggested by Gorbachev's glasnost model. I am contending that nuclear criticism is well situated to help understand this unlikely transformation.

One recent effort to find a context for the extraordinary trajectory of the Reagan era is the "structuralist and poststructuralist" approach taken by political scientist Diane Rubenstein. Her "presentation of Reagan as a hyperreal object" traces several instances in which critics expected depth or representation, and thus misunderstood Reagan, whose powerful effect relied on (and was unconstrained by) either depth or representation. Critics and apologists alike, trying to interpret Reagan gaffes, "simply do not realize that his episteme is not one of representation (in which the notion of 'false representation' makes sense) but of simulation, in which the [sign] of the real, the absolute fake is superior to the real itself." Critics were continually confronted with "obsessive and disturbing" reversals of their expectations, at the hands of "the Reagan thing . . . Reagan as obtuse meaning. . . . Reagan, the great communicator, leaves his critics speechless."[1]

Ronald Reagan's earliest attempts to deal with opponents of his nuclear policies rested on tried and true methods, which Michael Rogin has described as American countersubversion.[2] Accusing protesters of demonic motivations and methods, Reagan unleashed the rhetoric he surely felt had always been a cornerstone of his political success. Since the ear-

liest days of his political career in California, Reagan had appealed to a
constituency that was vague and undefined, except by its appreciation of
his charismatic rejection of protest of any sort. By 1981, when Reagan
gained a much larger audience, some of his supporters had adopted similar protest tactics, against abortion in particular. But, as Rubenstein
would surely understand, such oddities only drove Reaganism on. Reagan
steadfastly played his role, criticizing nuclear opponents in the strongest of
terms. His spectacular ability to defy conventional political lines infected a
wide variety of institutions and practices.

This odd inconsistency in Reagan's attitude toward protest could have
been interpreted as simply more evidence of laziness or lack of mental
ability, but such interpretations miss their mark by a dangerously wide
margin. Every presidential inconsistency could be reinterpreted as displaying the doublings and reversals characteristic of the manipulations
possible in this postmodern age. It even seems possible that Reagan's
popularity arose, in part, from his ability to phrase and direct our discontent with the technologies of political control and representation. Reagan
encased his code in a potent sermon against "guv'mint," especially
against its coercive component. Unwilling even to speak all the syllables
of the word "government," Reagan engaged (casually, it is true) the enterprise of living in the belly of the beast. In this sense, he practiced, for
all of us, the act of living amidst what makes one most anxious. Although
Reagan's anticoercive promises were obviously shortsighted and nostalgic, the myths he invoked set a context of doubles and literary creation,
the "construction of spectacle," in Murray Edelman's phrase.

It is a confirmation of this "strategy's" force that some of Reagan's
most outrageous statements received little attention. Even when Reagan
experimented with resolving the tensions of nuclearism through the mythos
of apocalyptic fundamentalism—invoking Armageddon—he generated surprisingly little attention and media coverage. The biblical apocalypse version
of the myth turned out to be just as reasonable as the mythical deterrent umbrella we had already found a way to tolerate. This example—and many
others—should focus our attention on interpretive analysis; the "concrete"
traces of interest and institution are immediately exhausted. Our collective
willingness to accept the looniest experiments at the White House is a dense
subject for interpretation. At some level, the press and public knew that
some different language game or political genre was upon us. No longer
were we involved in a game of interests and overt, legislative power that we
had always assumed constituted "politics."

Nuclear opponents responded to both the ideology of Reaganism and
its doubles by intensifying and broadening their activism. Emerging in the
early 1980s, the nuclear freeze signaled successful efforts by these groups,
at least for a while.[3] The sudden success of the nuclear freeze movement

easily outdistanced all previous "ban the bomb" or antinuclear power agitations. Mass demonstrations, ballot referenda in several states, and congressional action followed in rapid succession. Pointing to the new "conservatism" embodied in Reagan, this new movement quickly attained the status of the most important oppositional site. Activists mobilized citizens never before involved in antinuclear organizations and soon found widespread public support, as indicated by public opinion polls and other measures of sympathy.

While the burst of oppositional activity that culminated with the freeze could be read simply as a reaction to the wild Reagan rhetoric, there are subtleties to consider as well. In retrospect, now that we understand Reagan's importance as the first successful president of the postmodern age, we can recall the freeze as the first fair test of such a presidency. More than any other such movement, the freeze was as metaphorical and symbolic as the actor-president it opposed. The freeze movement signaled a shift when it convincingly presented itself as a contributor of alternative meanings, not merely of alternative policy positions on deterrence or a simple increase in the intensity of humanistic values. As I am suggesting in this chapter, the freeze partook in a new form of politics, even if it did so without much knowing it.

## Frozen

Freeze is, at first glance, a temperature metaphor, stopping the movement of something fluid by cooling it.[4] A "freezer," then, is an amplified refrigerator—an actor beyond the Cold War, with newfound abilities to make even the peskiest microbes (or the devils of nuclearism) *just stop.* The freeze emphasized its own modesty when it gently promised to cool the nuke, to make it less hot, less fearsome. The freeze thus carried with it a specific coding of nuclear technology—this technology threatens control, but it does so amidst a rhetoric of the audacious and unmanageable. The nuke encoded a truly tireless form of danger; without irony, one physicist called it "the catalytic nuclear burner—an inexhaustible source of energy."[5] At its most obvious level, then, the freeze staked out for itself the task of managing the wild.

Nuclear winter emerged, publicized by Carl Sagan and others, as another kind of temperature metaphor, warning of a freezing at higher, atmospheric levels. Nuclear winter aimed its findings at the nuke's metaphors, reversing the most primal consequences it might be imagined to have. The predictable outcome of nuclear war's explosive heat would be a resumed ice age, not an inexhaustible conflagration. The nuclear winter scientists even contested the nuke's coded desires—its unstated hopes for apocalyptic fireballs. It turns out that not only the operation of nuclear deterrence is paradoxical and inverted; its metaphors of fire and heat are also inverted. From the first appearance of "nuclear winter" arguments, we could no longer doubt that this political battle was a battle over codes.

The freeze had its own coded reversals and combinations. It issued signs of a generalized solidifying (stopping movement), but it also exposed the role of motion metaphors in this discourse. If the entire environment was now characterized by the integral role of speed and movement in warfare (as Derrida and Virilio also emphasized), every move—whether "accidental" or "intentional"—posed a special threat.[6] When the freeze emphasized that danger it leaped over the objects, the warheads that had previously been the exemplary center of these discussions, specifically in the call to "ban the bomb." Both in an arms race and in a war, speed (distance and time) becomes crucial. As Virilio claims, "In the current context, to disarm would thus mean first and foremost to decelerate, to defuse the race toward the end."[7] Encased in the freeze was the code that included this new context, even if the overt message was a (deceptively simple) "just stop here, now."

The freeze, in short, was more than simply "good political advertising," or rather, it was public relations in the best possible sense. "Freeze" is so multivalent that it points to *aporia*, the vertiginous gap one experiences when meanings unexpectedly conflict, double back, and confuse. The first sign of this politics, its odd and contradictory name—the freeze movement—itself signals this gap. Finding *aporia*, the freeze holds attention, in a way that problematizes the elements it encounters. For once, without resort to a mystified "reality" or comfortable, clear mythologies or ideologies, opponents had a way to evoke the gap—the void appropriate to their intervention into this odd public issue. The freeze was insatiable and shifty; victory in Congress was the occasion to problematize other arenas. The mood of that problematization could be in turn giddy and somber, reasoned and outrageous. No strong metaphysics underwrote this fractured (frozen) narrative.

We begin to engage the gap of meaning deconstructionists call the *aporia* when we recognize that "freeze" is not just a description—however multileveled—but is also a *command*. In a crime movie, the player with the gun (a criminal or a cop) hollers. In the case of the nuclear freezers, a reversal is at hand; the gunless victim is hollering, hinting that legions of reinforcements will be mobilized by the call, or perhaps reminding the bully of the consequences his actions carry. In this sense, the command— "freeze!"—takes charge in even the least likely circumstances. It is an attempt at reversal that focuses attention, and, one hopes, action, on the reversed term: namely, the relations of domination embedded in nuclearism.

More and more reverberations of the word "freeze" spin out, in the kind of list deconstructionists assemble to undermine the clarity of words we assume "mean something":

A "freeze-frame" in the cinema is the stoppage of motion in the most unlikely of circumstances. A still photograph is imposed on a motion picture; a thing that existed in order to move. The paradox of postnuclear military missions that accumulate hugely destructive weapons in order that they never be used is thus duplicated in the movies, a form that then gives SDI its Star Wars name.

Freeze can imply attraction or appreciation. When something is preferred, a person may "freeze onto it." That phrase recalls the suspicion (discussed in an earlier chapter) that our loathing of nukes also carries a lust to use the bomb, a combination also used in Timothy O'Brien's novel, *The Nuclear Age*.[8] But the phrase combines loathing and desire; when bare skin fastens to very cold metal it is "attracted," but the attraction is dangerous and frightening.

A "frieze" is an architectural feature, a horizontal band in which sculptures depict a scene. The most famous frieze decorates the Parthenon, a building completed just before the outbreak of the Peloponnesian War. It depicts many military scenes, in the context of religious procession.[9] Organizing civil life under the sign of military-religious interface, then elevating that sign to the top, is an old (and honored) pattern.

While stabilizing and solidifying, the freeze still carries with it the hope of liberation; we hope it *frees* us.

Freeze is so densely textual—so thoroughly evocative of language— that the domination implied by "unspeakability" begins to unravel. In other words, the ambivalence of a command that is also description, representation, and pun is not simply an instance of the ambiguity of language. This density is doubled again; the freeze targeted deterrence assumptions that themselves were also a "freeze." Jean Baudrillard's emphasis on the stability of this situation applies here. The ultimate end of politics, the final point from which all other politics will take its bearings, is itself a freeze:

> Wherever irreversible apparatuses of control are elaborated, wherever the notion of security becomes absolute, wherever the norm of security replaces the former arsenal of laws and violence (including war), the system of deterrence grows, and around it grows an historical, social and political desert. A huge involution makes every conflict, every opposition, every act of defiance contract in proportion to this blackmail which interrupts, neutralises and freezes them. . . . Energies freeze by their own fire power, they deter themselves.[10]

When one juxtaposes freezing metaphors with energy metaphors, deterrence can be revealed as a system of control, albeit a perverse and simulated control. Promising violence and practicing restraint, the binary oppositions of deterrence can produce paradoxical stability, even if that seems to be the opposite of its images. Within that context, however, the freeze movement still managed a distinct activism, matching the binary form of deterrence. It promised inaction ("Freeze!"), but then intervened broadly: first by revealing power, but also by destabilizing the paradoxes underlying this situation. The freeze may have appeared archaic when it implied that a speeding world should stop, just there. But it was hyperactive when it exposed paradoxes.

Ironically, the freeze also confirmed nuclear criticism when it was stymied by another literary device—a strange end for a component of a supposedly "unspeakable" topic. As is his now-familiar pattern, Ronald Reagan switched fables or, more precisely, added another layer of fabulousness. The Strategic Defense Initiative (SDI), which had almost universally seemed a crackpot scheme in its earlier, "High Frontier" stage, responded so well to the claims of freezers that it quickly subdued their movement. The "domination reversal" offered when the freeze demonstrated that simple, rhetorical adjustments might derail the arms race was, in turn, itself reversed. In a pinch, Reagan doubled back, as would the cinematic cowboy-heroes Reagan both worshipped and portrayed.[11] The most prominent gun advocate ever to be president, a man who rose from being felled by a handgun-toting would-be assassin only to wave off gun control as a fallen cowboy waves off the helpers who rush to him, Reagan invoked a bigger gun control.[12] The "umbrella" or "Astrodome" metaphors SDI supporters evoked were easily "disproved," but have been much more difficult to dislodge, perhaps because they mask a type of weaponry that is inevitable, even if it is not now possible. In short, the freeze was itself frozen, finally, by a move more mythic and literary than technological.

## Thawing and Other Instabilities

The paradoxical freeze is an attempt to make safe, but it is also an attempt to reallocate danger. This citizen movement creates citizens by reclaiming for them the ability to be dangerous, the potential to disrupt. Michael McCanles articulates the possible dangers involved in any deconstructive move against nuclearism. This age, it turns out, contains two equilibria, not one. While nearly everyone's attention focuses on a hardware balance, another literary (or textual) equilibrium also pertains. All parties must agree not "to launch a 'counterforce' attack of deconstructive . . . analysis against the other side's texts."

We dare not deconstruct the paradoxes of deterrence, because to do so is to leave ourselves naked of textual defenses. It is for this reason, I believe, that adherents of nuclear deterrence are made radically uncomfortable by advocates of nuclear freeze. The point is not that a freeze will leave us open to nuclear attack, but rather [that] arguments for a nuclear freeze derive from metatextual analysis of deterrence itself. [Such] analyses can, from the viewpoint of the players of the game, only unmask the paradoxes within their own strategies and leave them without threatening force.[13]

McCanles is willing to accept this increased danger, finding it preferable to the false and oppressive stability of the existing paradox. But our understanding of this entropy does have effects. As a consequence, acknowledged somewhat obliquely by McCanles, the nuclear debate is politicized. The folly of bipartisanship—the myth that "everyone agrees" on this issue—is dispelled. Politics proceeds (in fact, it is intensified) on the grounds of language.

At the simplest strategic level, "freeze" is a double message, both negative and positive. While the negation of nukes still pervaded the opponents' strategies, the freeze label also implied a response, an activity. Obvious as it is, this puts the freeze into a different context. All other opposition had called citizens to their issues on the basis of fear of the nuke's grotesque potential consequences. When the freeze reallocated danger, it reclaimed arenas for action. It insisted that if control systems are going to prevail (as seems inevitable, given the generalized danger of the situation), that system will at least have to include players outside the Kremlin and White House, namely, the subjects of that control. Gorbachev recognized this consequence earlier than the Americans; much of his attractiveness, early on, surely rested on a recognition that he, at least, could acknowledge popular arms control claims for what they were—that is, assertions that the rhetoric of deterrence could still be deconstructed, and that this deconstruction would open political space, not end our prospects. The deconstruction of the Cold War balance of terror was not necessarily the same thing as the end of the world.

In this view, the simplicity, symmetry, and activism of the freeze never needed to reenact the fearful entrepreneurialism of disaster I discussed in an earlier chapter. The freeze edged toward the aesthetic, rather than the economic, exposing (if inadvertently and ambivalently) the political efficacy of alternative political approaches. We begin to see that political action could take place (and has already occurred) on a different basis than survival or unspeakability. The endless calculations of apocalypse—will I be safer in Montana than in California?—can be displaced by a simple, uncalibrated, resolute, and popular command: "Freeze!" The implication of that escape from calculation and calibration is that, even at this late date, politics still might emerge from a different place, from bases better inscribed by the expressive "individualism" identified by Kateb or the "aesthetics of existence" Foucault identified in his last interviews and books.

The response eventually generated by the freeze confirmed this reading. Reagan's Strategic Defense Initiative (SDI) performed a deconstruction on contemporary weapons politics, reacting more to a reading of antinuke narratives and rhetoric than to any specific technological invention. Although Virilio's *Speed and Politics* preceded SDI (not to mention the Gulf War video show of 1991), the automated, light-speed character of Star Wars is precisely foretold in that narrative of the compression of space and time. This, now, is the way war passes entirely beyond human agency, matching the movement of power away from intentions and conscious plan. As Virilio notes, "The possibilities for properly human action will disappear in a 'State of Emergency.' [Telephone] communication between statesmen will stop, probably in favor of an interconnection of computer systems, modern calculators of strategy and, consequently, of politics."[14] By the time the buildup began in Iraq in the summer of 1990, the *Wall Street Journal* could brag unabashedly at this capability: "In many cases, these advanced systems would make many of the judgments about who the enemy is and when to fire at him."[15]

This pace seems to represent assurance, but this is always paradoxical. Pace no longer represents competence; now, it is a reversal. New destabilizers constantly emerge to confound the stability of the nuclear age. The signs of safety continually appear as accompaniments of chaos, and the inherently chaotic (the nuke) raises possibilities of a more managed society than anyone had ever imagined. Reagan's well-known inability to understand how our "defensive" capabilities could appear obviously "offensive" to the Soviets is only a symptom of a larger tendency that pervades nuclearism.[16] As Garry Wills has explained, resources will inevitably be confused with intentions; resources become the sign of intentions, and the reality of the sign is continually overestimated, perhaps more so when the stakes are higher.

> Naturally, the enemy's intent and willpower are less visible than his resources; so we overestimate them in much larger degree—this is called the "worst-case" scenario. If we must presume the worst in order to be prepared for anything, then the slightest increase in enemy resources must be read as part of a larger design being implemented. Even a cutback in one area will be read as an economy called for by greater expenditure elsewhere.[17]

Transposed into the reverse logic of deterrence, the consequence is that assessments of enemy strength—a more or less routine affair in peacetime—become permanent destabilizers when the balance of terror is institutionalized. The rationalistic management that modern nuclearists proclaim as their achievement will continually threaten to produce aggression and unbalanced terror. In such a strange setting, as Deleuze and Guattari explain, desire will stage breakouts along all sorts of unexpected lines. The robotic displacement of human bodies is one such site. As I suggested in an earlier chapter, robots are one of the key devices by which we impart a meaning to otherwise opaque, elusive, confusing, and destabilizing technologies, bestowing form, or presence, in space. We are now ready to return to the robot, which played such an important role in the SDI debate.

## C3PO, R2D2, and Darth

The most popular instances of such a "solution" are the friendly, subservient cyborgs featured in *Star Wars*, the movie that became an anti-weapon weapon.[18] The *Star Wars* movies are, in part, fables of human control over technology; the robots C3PO and R2D2 are the stars at war with the dark forces. This, I would argue, is where the most serious conservatism of these movies is lodged, more than in their stereotypes and predictable Manichean plots. These latter aspects of the movies were most often the objects of critique, no doubt because we are so familiar with a critical discourse that calculates and balances the reified signs embodied by stereotypes.

To be sure, it could be that Reagan returned so often to *Star Wars* terms precisely because that movie symbolized the struggle of light against darkness, good against evil (empire), us against them. I am only suggesting that another reading is available. The *Star Wars* films did something more important than simply reinforce the practice of identifying an evil other and then using that role to establish popular support for military response. Put simply, the movies also did revisionist ontology on the mechanical man. Robots such as these *Star Wars* characters remind us that we are in control, no matter how exotic the technology; in this century especially, we know bodies (such as the ones persons, if not bombs, have) can be subdued.

For once, the more the robot characters seemed human, the *less* threatening they were. Their similarity to us does not represent dangerous autonomy; we aren't dangerously autonomous. Reagan picked up on precisely this aspect of *Star Wars* with his use of the saga. After all, it was always a central part of the Reagan rhetoric to persuade victims to participate (if only by proxy) in their own oppression. Reagan's Star Wars, then, pits our good, obedient robots against the destabilizing, anarchistic mutants (Darth Vader, the quasi-android Andropov of the Evil Empire). The robot has been welcomed back into our brand new, all-electric, General Electric house.

The *Star Wars* movies always worked best in the realm of symbols and intertextuality. That is to say, they interconnected with any number of genre movies, from *High Noon* to *Triumph of the Will*. *Star Wars* is so flamboyantly about the codes of genre that it surpasses other Hollywood movies in a giddy display of code and decode. No longer nervous about its metaphors, *Star Wars* can stretch them so wildly that they become the fun. In one crucial example, R2D2 is the child in a surprisingly childless children's movie; short and inarticulate, it is nonetheless indispensable, in part because of its loyalty.[19] The boy doesn't have a dog, but there is a Wookie, likewise inarticulate but faithful, strong, and competent. The seemingly untranslatable mysteries—a Wookie's bark, R2's bleeps, or the obscure goings-on at an interplanetary tavern—become signals that can be understood. Dealing in the intertext, then, becomes the parallel for any child's "coming to terms" with the confusing, often disturbing adult world.

*Star Wars'* strength derives from this intertextuality; a remote, highly technological society turns out to be utterly familiar. There is a sense in which that familiarity makes the film conservative; youthful rebellion, yankee ingenuity, and arrogant appropriation must be natural to humanity. Perhaps it is even hyper-conservative; in a crucial scene, diverse species from throughout the universe commingle in that tavern. They misbehave; after all, these are the most innovative scoundrels of every species. But they are subject to control. All of the *Star Wars* films are mastery fables at their most obvious level, but this is also their least interesting aspect. It is as if all this mastery is planted in the movies to distract liberal critics from noticing the important reversal that is the series' real development; the new robot was now possible—self-disciplined, as cynical (and hence useful) as any conventional human being.

Still, the films are not as fully and ideologically conservative as critics often conclude. Amidst the heroism and paternalistic sexism, subversive themes persist. Within the films' scenarios, technology is undermined, specially planned so as to be dysfunctional; that is, after all, implied by the new character of the robots. Reviewers of the film noticed the beat-up, spaceworn appearance of Han Solo's spaceship, but may not have understood the role it played. This is more than homage to Gabby Hayes's dilapidated jeep in the old Roy Rogers films. At first glance, the seemingly decrepit Falcon could be a sign of cunning; after all, Han Solo's (the *solo*, solitary *H*-um-*an*) warship proves itself fully competent. Nonetheless, doubt persists; is this persistent machine-cough a sign of cunning, or of planned obsolescence, the consumerist version of Nietzsche's struggle with death? How can we answer this terribly important question, without using these machines of destruction—exactly the experiment we wish to avoid? What difference does it make if these bumbling machines *don't* work?

Just as we love and hate the new technological agents we find in our midst, the *Star Wars* films vacillate on their judgment. The theme evokes mastery and triumph, but the signs always threaten to reverse themselves. The entire series of films (promised in the very first moments of the first installment) encases itself in fatalism—the tale presumably happened "a long time ago." We can only presume that the technology failed its masters. (This feature of the films also evokes themes of history and memory. Films have become aware that they are a sort of collective memory of something that never happened, recalling Fredric Jameson's notion of an "identical copy for which no original has ever existed."[20])

What finally solves this indeterminacy in the movie is the escape to farther stars, the transformation of machine matter into light and speed. Over and over, the heavy, metallic vestiges of the nineteenth century succumb to the light-speed tricks of the future. The blast into "hyperspace" may be scientifically dubious, but metaphorically it is right on track; as much as the Industrial Revolution transformed the world, the Communication Revolution (using Haraway's light and Virilio's pace, not Marx's machines and power) will eclipse even that change. In the *Star Wars* films, the signal of that change is the laser sword. In place of steel, it uses light— glowing, quick, portable, and competent absence. Armor resists this new weapon, but only for a while. The light wins out.

Star Wars technology—the Defense Department's, that is, not movie director George Lucas's—elaborates the themes of lightness and absence announced earlier by the *Star Wars* movies. Major components of SDI would be in space, away from any earth base, and would neither carry fissionable or fusionable material nor fire traditional shells. Instead, major parts of Star Wars reflect beams and sling nonexplosive "kinetic projectiles" (essentially, space nails). Critics helped publicize the importance of huge computer programs (and programs to test programs) for the success of the Star Wars project. The abstract control technology that failed at Three Mile Island would have to be perfected in a far more dangerous setting; this war would have no observable battlefield, only screens and instruments. When the technicians in the movie *China Syndrome* encountered implausible readings on their gauges and screens, they responded by tapping the face of the gauge. Their entirely human response is the sign of our future; we stare in disbelief at the deranged indicators, tapping the gauges, as if to bring them to their senses. Or perhaps we are asking them to let us in, as if we were tapping at the door of a new world. Absence, abstraction, and exclusion pile up on one another in this brave new place.

That Star Wars obscures the apocalyptic consequence of contemporary warfare only adds layers of absence. The rhetoric of Star Wars differs from all other military rhetorics of our age, explicitly assuring us that there now is a defense against this (absent and present) nuclear threat, about which we have only known for certain that there was no defense, and that any possible defense would be unimaginably destabilizing. The defense, itself, will be invisible (located in space, and relying on components which could not be directly viewed), creating more absences. The need for instantaneous response is so great in SDI that the decision to go to war—one of the central decisions any political system before now has made—will have to be automated. That moment of political life when the most grave decisions are made—to declare war or not—has already been displaced from legislative control. Now it passes from direct White House or Pentagon control, as well; that solemn power necessarily escapes gravity and disappears.[21]

Although Lucas steadfastly explains his films as sociological texts, *Star Wars* had a subtext, welcoming the friendly robot and replacing politics. Before, cyborgs were either seemingly friendly but ultimately demented (notably HAL—IBM, one letter displaced at each initial—in *2001: A Space Odyssey*), or competent but purposefully disabled in one human attribute or another (notably the only incidentally biological Spock in "Star Trek"). With *Star Wars*, the robots moved in fully. The only disabilities R2D2 suffered were that he was somewhat short (an asset, it turns out) and hard to translate; those disabilities only indicate a residue of cunning and humor in his design, translated as self-assured confidence. Dependable, independent, and loyal, robots could now stand in for humans in all respects. This is the fable, or mythical apologia, for the true role of the robot, which is to finish the demise of politics that has been on the agenda of technology throughout the era of modernism. The robot *must* do what has been politics; no choice remains. Star Wars was the perfect name for the Strategic Defense Initiative (SDI). That may explain why this nickname was so easily accepted by media and public, compared to Reagan's prolonged (and generally unsuccessful) campaign to rename the MX missile the "Peacekeeper." SDI is to the nuclear threat as R2D2 is to threatening automation. With help from SDI, the bomb now wears a happy face and beeps like a Nintendo game.

Indeed, SDI has been resistant to criticism, despite the almost giddy claims its supporters have made for a system that pushes our technology well beyond familiar understandings of its limits. The deterrence logic remains; SDI would not actually have to *work* to be effective. If Star Wars doesn't work, it doesn't matter. What must work is our national pose as *programmers superb* or, as computerists might put it, as "insanely great code writers." SDI's cost doesn't matter, since there is no market value on its purported product, survival. To debunk SDI, opponents must presume that it would be used, and no argument with that premise can succeed. SDI's critics, who found themselves strangely neutralized at the symbolic level of nuclear discourse, were surprisingly vulnerable to Reagan's fanciful arguments. Reagan's advantage had always been that his facts are irrelevant, and therefore do not represent as serious a test as Joe Biden's and Pat Robertson's facts. What is compelling about Reagan, as Garry Wills and Michael Rogin have shown, is his ability to live in the world of simulations and stand ins.

This is another "giddy incompetence," this time achieved at the very center of politics between the major nuclear powers. Competence—the human talents arrayed in projects—is irrelevant. The oblivious giddiness of a Reagan blithely replaces grim realism, substituting a television commercial in its place. In one laughable advertisement placed by a private organization, cartoon rockets ping off a protective "Astrodome" shell. Beyond the Big Lie, this wild misrepresentation requires no hidden truth—it is absurd on its face, not even attempting to either deceive or represent. The "Astrodome" advertisement was so outrageous it was able to present itself as a child's dream without undermining its images of competence.

Put another way, the Strategic Defense Initiative was and is a literary invention, a fiction midwifed by R2D2. SDI was immediately successful in its Star Wars variant, even though no technological "breakthroughs" provided grounds for such optimism. But "real" technology was almost irrelevant in the face of a masterful fiction that worked both sides of its street. Arms control talks stalled on the issue for a half decade, and the nuclear freeze, which had seemed such an apt response to the bomb, was quickly subdued. These failures and multiplicities mask a type of weaponry that is inevitable, and whose presence is now underwritten by that other inevitable, metaphorical machine, the friendly robot.

This formulation—the friendly robot—challenges humanism as much as the cartoon figure Casper the friendly ghost repudiated death (while also, like the *Star Wars* robots, flattening the prevailing mythology onto a postmodern screen). Accordingly, it should not surprise us that the humanistic critique has displayed increasing anxiety as we have become better acquainted with all these technologies. The facile "science-fair nerd" model of computing could hardly survive once so many of us had experienced that engagement. It is no longer "the other" who has been captured; instead there is now an other—a cyborg—that threatens us with capture, but also invites us to dance. Obviously, that captivity needs to be clarified. The humanistic tendency to lodge this encounter with potent machines in an "addiction" discourse is easily deconstructed, as I suggested in an earlier chapter.

## All That Is Solid . . .

One lesson (increasingly noted by both nuclearists and their critics) is that there is very little that is dependably "human" for us to rely upon in formulating responses. Star Wars points toward a broader role that the robot must play in this discussion when it begins to take over every role of warmaking: surveillance, detection, identification, decision, and reaction. It is the robot that finally delivers us entirely into the realm of simulation, displacing old structuralisms with such finality that "poststructuralism" it-self seems not only an awkward label but a nostalgic one as well. The robot severs the line back to sociology. It cannot be an analogy or stereotype, since it displaces the human referent, exchanging it for artificiality. The phrase, "stereotypical robot," is a senseless duplication; it is the robot that stereo-types its creators, but the fact that the robot is both so necessary and so ex-emplary makes that attempt to domesticate it merely hyperbolic.

While the robot displaces the body, the ego is displaced by velocity, the speed Virilio emphasized. For political opponents to ask that these meth-ods of warfare "just slow down" is to ask the impossible. (This preemi-nence of pace was vindicated in 1989–90; the geographical division of Germany—the ultimate statement of the importance of territory—became much less important, while the turf of the Middle East, of Iraq, Kuwait, and Saudi Arabia, became much more important, since it is a crucial source of oil, which still fuels movement in our age.) In a general, political sense, antinuclearists confronted a politics that had awesome abilities for slipping away from criticism, since that politics no longer conformed to existing presumptions. With the simulation and displace-ments firmly in place, every critique was undermined. Calls to stability met deterrence responses. Calls to tolerate instability met fear of the nuke.

In view of nuclear criticism, a more complete reevaluation of the po-litical demise of the freeze is now possible. Tracing the freeze's political trajectory, we easily discovered that the movement frequently fought the wrong political battles, making Reagan's counteroffensive simpler than it should have been. Freeze activists held bitter internal debates about their position on the "moderate-to-radical" continuum. The movement, wish-ing to maintain a moderate stance, suffered debilitating internal fissures as it replaced veteran activists with new organizers, chosen primarily to represent "actual citizens," thereby avoiding the strident radicalism of the past. The tragedy of these battles is that the freeze had already moved beyond the discursive context within which the kind of debate they con-ducted is lodged.

The aggressively multileveled discourse the freeze opened, a way of talking that insisted on a stoppage that was simultaneously progressive, had already dealt with the need for such moderation. Having skipped over the old political space of populism and Marxism, the freezers wrongly insisted we go back and visit it again. Reagan's adventure into yet more dizzying literary space, into Star Wars, ended the freeze by highlighting its own conservatism. Well trained in the movies and in media politics, Reagan just knew the shape of postmodern politics. Having stumbled upon a new political discourse (in the same manner poststructuralists describe authorship in general), the freezers never knew where they were.

Antinuclearists made matters worse, then, by radically distrusting the engagement, the level of fascination required by all this new technology. The new ensemble of techniques—computers, robots, bombs, and their associated human disciplines—is so completely involved with human images, thought, work, desire, and language that it forms a new agent. No longer a "mere tool," technology now positively—and rapidly—transforms human subjects. The most obvious illustration of that changed relationship (not a sign, in view of the change, but a mirror) is how thoroughly this technology captivates—holds captive—our attention. In ways even Alice never imagined, these mirrors absorb us.

When the movies cross paths with such an integral center of our daily fascinations, our mythologies, and our most demanding discursive puzzlements, they establish a site where political speech can attain a special competence. Any language-and-politics position must recognize the utility of literary production in politics, and that is exactly what Star Wars exemplifies.[22] With Ronald Reagan's finely honed ability to deal with images and fables, Star Wars quickly became the solution to the challenge offered by the freeze. Light (lasers), speed (computers), and the openness of space promised a release from the machine world of tanks and bombers. Of course, Reagan hedged his bets with such attempts at hybridization as the Bradley fighting vehicle and the B-1 bomber, but these half measures suffered predictable difficulties. The real logic of Star Wars was a new fugue on deterrence. If deterrence worked by promising future, fictional destruction, Star Wars promised peace by predicting future, fictional defenses. This "solution" was entirely literary, of course, which is why the Star Wars name is so appropriate.

The critique of SDI has been so thoroughly rehearsed that I need not survey it here. As the Reagan era ended, Congress reduced SDI funding and it seems unlikely that anything resembling SDI will be implemented in the near future, even though this hyperspace overcoming of deterrence theory's discomforts seems likely to enter permanently the discourse of nuclearism. At the very least—in yet another reversal—SDI served to indicate an important political development of the era. Gorbachev, who may have understood the political fabulism behind SDI proposals, finally just gave up his objections to SDI in arms control negotiations. What had served as the major block to progress in the 1980s just disappeared, when Gorbachev simply discarded what had appeared as a political obsession. Thus, SDI, which had aimed a literary missive (missile) against the freeze — itself a concentrated literary site, a hardened silo of opposition — was itself, in turn, pushed aside by Gorbachev, who understood how crucial to politics this ability to change directions, seemingly without "objective" reason, could be.

Still, SDI produced effects. The freeze was instantly displaced, or thawed. That is not something to mourn as much as it is a characteristic of postmodern politics; disengaged from the real, politics will flit from one fascination to the next. This is not necessarily the end of politics; even Machiavelli understood the utility of changing political directions and commitments.[23] Modernism first identified this break, as Marshall Berman has explained. Marx's notion of revolution meant that change had become liberated from earlier rhythms and patterns. "The intense and relentless pressure to revolutionize production is bound to spill over and transform what Marx calls 'conditions of production.' "[24] Berman locates a "great imaginative leap" made by Marx when he was "propelled by the desperate dynamism he is striving to grasp." The site of that leap is the *Communist Manifesto*:

> Constant revolutionizing of production, uninterrupted disturbance of all social relations, everlasting uncertainty and agitation, distinguish the bourgeois epoch from all earlier times. All fixed, fast-frozen relationships, with their train of venerable ideas and opinions, are swept away, all new-formed ones become obsolete before they can ossify. All that is solid melts into air, all that is holy is profaned, and men at last are forced to face with sober senses the real conditions of their lives and their relations with their fellow men.[25]

But, as Baudrillard has explained, when the pace of transformation disconnects from the real, we enter a postmodern age in which the movement becomes frantic, even hyper. This is a change of category, a transformation into hyperreality that makes Berman's humanism seem quaint and nostalgic. To attempt reclaiming human pace, or turf, is to play a game that has ended. Deterrence has already deterritorialized politics, with both images (hyperreality) and speed, which is precisely important because it obliterates the territory of the battle.

Taking Baudrillard seriously, we might guess that the effects of the flash of opposition between the freeze and SDI actually spread out throughout the political world, instantly creating change where none seemed possible. Even the ascendency of Gorbachev — the U.S.S.R.'s first postmodernist, their answer to Reagan in more ways than one — might turn out, later, to have been one more consequence of that flash. The failure of the freeze, its inability to hold the ground it had claimed, put the opposition to Reagan and nuclearism back on familiar turf. That response — the subject of the next chapter — only seemed closely related to the freeze. As it turned out, the recourse to "lifestyle" was disastrously apolitical, sending old freezers out to chart the emergence of Gorbachev and the other developments of 1989. If only for a brief, mistaken moment, those events seemed not only fortuitous, but positively *saving*.

# 5

# Immodest Modesty

In an interview, Michel Foucault once noted that the local and partial interventions (from what we might call the left opposition) have succeeded more frequently, over the last several decades, than have more global and ideological approaches.[1] By "local and partial," Foucault recalled issues confronted on some personal (rather than universal or theoretical) basis—even if the issue has a global, foreign-policy dimension. In this chapter, I want to consider what this insight might say about the politics surrounding foreign policy in the advanced media culture of the United States.

To address this question, I will generally assume the (admittedly arguable) conclusion that localism has been effective. Reactions to U.S. policy in Vietnam and Central America, in arms control negotiations, and to a lesser extent on the Palestinian question should serve, I think, as at least a sufficient list to suggest the plausibility of what follows. A more precise statement of my question thus would be as follows:

Insofar as localist initiatives have been successful in foreign policy debates, what is the political logic through which they appear?

Is there reason to suspect that disjointed, ironical, and incomplete claims of some particular sort may have a special status "in spite of" their obvious liabilities within dominant ways of addressing these questions?

To narrow this discussion once more, I will consider one particular localist approach: the commitment to intensely private, personal practices in

86

the name of (and presumably somehow linked to) larger issues of world peace. Along with the freeze movement discussed in the previous chapter, this "personal" politics composed a distinctive 1980s oppositional style. For many Americans, this became the only legitimate response to the absurdities and horrors of the nuclear age.

To engage this topic is to venture away from the usual patterns for political analysis of foreign policy debates. It is to take as crucial that which the dominant discourses exclude, and to pursue in those exclusions the traces of power that ordinarily remain unproblematic. In short, I would follow R. B. J. Walker's advice, and refuse to privilege those actors who have constructed for themselves a central position.[2] Even in the discourse about foreign policy that has so privileged its central actors, a mode of analysis that seeks to politicize will have to contest—but surely not ignore—that privilege. This approach explicitly forges a move toward the margins. Up to now, most serious discussions of such politics have confined themselves to the (more or less) coherent positions taken by elites.[3] My study does not address those approaches directly so much as it displaces their assumed realism (and their consequent psychologizing of this politics).[4]

I would not be doing excessive violence to the many studies done on the topic of attitudes about international politics if I were to misread them, finding a subtext, the record of a scattering, a divergence. In other words, the coherent and thorough discussion of elite foreign policy attitudes that has proceeded for years could be reread as having had an inadvertently deconstructive outcome. Seeking to identify positions with those who hold them, these studies have shown instead how difficult such an identification would be. Despite concerted research efforts, scholars have failed to settle on the categories into which even this small, homogeneous group's attitudes fall. Repeatedly, the reference (a summarized position) slips from its referent (politics). Wittkopf's finding is, thus, one in a series.

> Previous analyses of the mass surveys . . . demonstrate that a single
> internationalism-isolationism dimension of the sort implied by the
> internationalist consensus of the Cold War years no longer adequately
> captures Americans' attitudes. . . . Instead, two dimensions are required.[5]

Or perhaps three[6] or four.[7] Or perhaps we need to include "the nature of power in America" in our scope, as a condition—more than a dimension—of inquiry. Perhaps "the whole system of thought [has moved] to the right."[8]

In this chapter, I wish to train some alternative instruments, rather than filling in new dimensions and conditions by conducting surveys of elites.[9] The "instrument" I have in mind is language. In brief, I want to take the tropes—rather than professions of tropes—as the data, as prior to any questionnaires we might distribute. My focus, with this distinctive instrument, is also alternative. My marginal, fragile approach would pursue a similarly flimsy object: the anti-nuclearists, whose elite and popular practices of opposition help form the context for foreign policy debates. Leaving the surveys behind—or rather, starting just before them—I would study these oppositional stances by examining the rhetorical contexts, the patterns of language through which these critical positions emerge and assert themselves.

## Lifestyle Strategies

The point of opposition I wish to analyze (to which I will refer by the ambivalent name "lifestyle politics") arises from an observation regarding politics in the contemporary era; the *self* somehow has become crucial to every political critique. Each movement now claims that one's ability to reformulate one's own actions is the key to *political* change. The world of political possibilities somehow aggregates from individual, conscious choice. This connection has informed friendship tours of Americans to the U.S.S.R., the Great Peace March, and a whole range of rhetorically "peaceful," noncombative ways of life. The slogan "Think globally, act locally" signifies this approach.[10] With one eye on global, ideological meaning, activists move toward utopia indirectly, by expressively redesigning the ordinary matters of life. That is what recycling efforts and the natural food movement both promise. It is the model for the critical pun, seen on automobile bumper stickers: "You can't hug a child with nuclear arms," or "Think Peace" (a reference to an old ski area slogan, "Think Snow").[11] Perhaps the most poignant—"Be all you can be; Work for Peace"—brings together the intense concern for self with a reversal of military advertising, itself a form unknown or completely awkward until recent years. All of these disciplines of a modest personal style have an immodest goal: to reconstruct world politics, here at my specific compost heap.[12]

## The Social

The term "lifestyle" almost automatically evokes ridicule. Overtones of commodity fetishism and intensely personal, stylized standards of meaning are reminders of the yuppie, everyone's favorite scapegoat of the 1980s. Some of the practices of the antinuclear movement are indeed trivial, hard to take seriously. The slogans mentioned earlier are tiny, cute interventions that belittle the importance of their topic without raising their claims to the level of irony, a potential move I will discuss in a later section. Struggles with such intensely symbolic activities as personal solar heat technologies seem a diversion rather than a problematization. A thousand flowers bloom with the same characteristic petals: self-absorption such as Werner Erhart's Hunger Project promise world peace based on personal behavior, but produce (literally) precious little. Perhaps most clearly, the move to the countryside is a move away from the city (where problems concentrate and come into focus, or at least into heightened contradiction) to a simpler—that is, unproblematized—place. The "tracts" are, indeed, sub-urban.

Still, these frivolities can be put into service as reminders. They form a politically important practice, even if these "American Greens" face structural barriers to the types of participation found by their European counterparts.[13] One even hears, among Americans, that it is their responsibility to "get their own act together" before entering politics. That is to say, lifestyle attains a prior precondition for political activism, providing a standard by which the foibles of activists can be criticized without dealing with the political claims they make. No doubt, American apprehension about politics in general also informs this sort of precondition. Encompassing critique, tradition, and daily habit in a unified set of practices and attitudes, these lifestyle disciplines resonate, despite their obvious limits.

These commitments reside at an important place in our political landscape. On the one hand, this plea to (even mundane) practice is in some sense new to oppositional political stances. These practices presume that the *self* is at stake, and that how we actively and consciously reformulate ourselves (through our habits and daily practices) *ought* to have something to do with this politics. On the other hand, there has been a long tradition in Western political thought connecting the "personal" to the "political." Following the classical liberal era, the idea that individual behavior produces the basis for a moral order gained a special status. Society came to be understood as composed of aggregated individual actions. Clearly, our democratic traditions trace back to a strong version of this presumption. Nonetheless, the current claims are at least subtly distinct from these older forms. The meaning of "the self" has not remained unchanged throughout this century of gathering professionalism, therapy, treatment, and education. We should remember that even before this century, serious, repeated disruptions threatened the connection between the personal and the political.

Marx's disjunction is the most famous; he decisively severed the way we needed to act, to bring on a new order, from the way we might act once that order came into being (or, for that matter, the way we might wish to act in our humble, gentle everyday encounters). After Marx there could be no mistaking the presence of this coded disconnection at the very core of the dialectic—awareness of the disruption becomes the standard for assessing all modes of understanding politics. This was not an isolated development. Adam Smith noted that acquisitive, aggressively private behavior could produce a *moral* order. Marx redoubled this severance when he explained that commodities and our fetishes about them have indirect, structural meanings. After he addressed the connection between personal choices and political meaning, that relation was deeply problematical.

In recent decades, the connection to future order is even more thoroughly severed. After Foucault, Marx's notion of indirect, dialectical progress toward moral order is more a part of the liberal context than a refutation of it. The "self-satisfied authority of the humanist voice" has derived from sources now in question.[14] Baudrillard expressed the limits of the Marxian dialectic, as well as humanist forms, when he explained that the social had "imploded." The social has not been "overcome" (in either Hegel's sense or King's); something beyond concept and term has happened to this central ground of the dialectic. In Baudrillard's extension of Foucault, the notion of society has been durable because it has produced the "effects of truth." But this durability does not denote a significance that goes any deeper than its own constructions. In other words, the underlying metaphor of society might not be the production of meaning, but magic, symbolism, irrationality, or defiance.[15] In short, our understanding of the social may lack imagination.

Baudrillard's idea of the social—some commons of thought, action, coercion, and negotiation—is distinctive, clearly different from that presumed by any citizen attitude study. On one hand, the social is deconstructed when it is displaced by "the masses," a floating referent that could point to philosophical or physical substance, to a group, or even, in French, to geophysical or electrical properties.[16] But, on the other hand, the mass can be subjected to commentary, an opportunity opened, for example, by Nixonian references to a "silent majority." In either case, "the mass is what remains when the social has been completely removed."[17] Any opportunity for action that might remain would no longer be informed by a society that understands itself to be a society, capable of concerted action.

If the social has been removed—or if it never existed—what we have called "sociality" may actually be its simulation. Without a social model, we could no longer take for granted our assumption that the actions of a group or individual "reflect" or "represent" a position. In the realm of simulation—in the simulacrum—the terms usually attached to a politics of significance (of signification) can reverse. Even more dramatically, these terms can disappear. In an important example of these effects, the apathy of the contemporary mass is, in Baudrillard's telling, an absorbing, unreflective "black hole." But that indifference is also a positive development, a sort of strategic refusal, the masses' "true, their only practice."[18] If the masses have escaped signification and its political form, representation, this does not necessarily mean that they have "failed."

> As Nietzsche well knew, it is in this disregard for a social, psychological, historical truth, in this exercise of simulacra as such, that the maximum of political energy is found, where the political is a game and is not yet given a reason.[19]

Seen this way, the indifference of the masses escapes a politics that has been "dominated by representative mechanisms" for three centuries. Marxism suggested that the masses would represent social fact to political institutions, eventually dissolving the political, withering it away. In this sense, it may not be so strange to understand that the social is what disappeared instead.[20] And without a signifiable "social," any politics based on representation begins to falter.

Marx also argued that the negation of the social—alienation— confirmed the social possibility, but Baudrillard closes that escape route by, in effect, deconstructing alienation itself.[21] The silent majorities still speak, but refuse to be represented. "And in this sense, far from being a form of alienation, [this silence] is an absolute weapon. . . . This is its mode of defense, its particular mode of retaliation."[22] Still, Baudrillard is continually evasive on diagnosis. These strategies are dangerous, they require energy to maintain, but they also present opportunities for intervention. The masses join the media and terrorism in a "triangular affinity,"[23] but that affinity may only help us account for developments that otherwise seem merely brutal.

This interpretation of indifference and apathy seems almost automatically to produce annoyance among American intellectuals; after all, so many of their analyses now rely on the conscious, calculating, reasoning individual. This deconstruction will have to be more than simply striking. The intellectual enterprise itself would seem to be threatened with extinction if it could not depend on the presence of calculating, participating individuals-with-attitudes who respond to information. As if to acknowledge what is at stake when he courts the apathetic posture, Baudrillard finds one of the best examples of this implosive inertia in immediate proximity to the locale where most American intellectuals work: on the college campus, in classrooms, among their students.[24]

With rare (and often regretted, as in the infamous case of Allan Bloom) exceptions, student reaction to their own campus politics is the purest black hole. This reaction is always interpreted and deplored as "student apathy," a stubbornness to react to all those wonderful lectures and readings, even if these performances have yet to reassure a student body obviously terrified by their (just as obviously) diminishing prospects. On one level, such an analysis is a psychologizing tactic; the student psyche is to blame. But there is a different way to see it. Student politics, much of the time, seems to most students to be a "sucker's game," which one would engage *only* for psychological reasons, not to address power. In a sucker's game, there is some double bind arrayed, in place of chances to win. Most of the time, students are not able to negotiate with the people who matter most, in the scenes of power. Students don't set most of the taxes or many of the rules, even if the institutions of their governance *seem* governmental. There isn't even a police force.

Students react by going inert. They become a mass, silent. Those who have power over students insist that the wages of representation are responsibilities. An acute student recognizes that the opposite is true; only "irresponsible" behavior produces a real chance to negotiate with those in power. The contradiction means that all utterances about "the students" as a political force on campus will necessarily be metonymic. If a "mature student body" is mentioned, the coded reference is surely to large, always "enthusiastic," always infantilized rooting sections. Surely this irony is not lost on most students; embedded in an institution ostensibly dedicated to "higher" levels of everything—especially maturity and erudition—they discover that the standard by which the institution wishes to judge them is how well they have reverted to the old "school spirit" of their high school days.

Given the craziness of the game, as students might put it, they go inert, in just the same way as Baudrillard shows the public reacts to public opinion polls. Both "groups" refuse to be characterized, or at least refuse to participate by responding in the same spirit invited by those in power. Students refuse to participate in student government, and seldom speak of it seriously as a representative mechanism. Student elections continually evoke fake or cynical or self-consciously "just plain dumb" candidates with platforms to match; pranks abound, clearly denoting a refusal to take the charade seriously. For their part, the general public—some of whom may have learned this in college, as the saying goes—starts refusing to play the straight man to the pollster, leading pollsters and social science analysts to conclude that they are ill informed. We could argue, with just as much evidence, that their disappearance is a strategy, a subtle revenge.

This demise of representation, its removal from the center of all political acts, is an important development, but it does not necessarily end politics. An expressive practice could deny representation, could intervene in politics instead through disruption and ironic juxtaposition. In our present case, we might begin to wonder whether lifestyle expressions *represent* a preferred future (as they sometimes claim). Perhaps, instead, the whole system of solemn, serious expression of political positions is somehow undermined by these solemn and serious substitute lifestyles.

## Language and Political Strategy

Once we notice the crisis of representation and its first consequence — the demise of the social throughout the theoretical landscape — other questions arise. Returning to the questions posed at the start of the chapter, I can now suggest that insofar as localist interventions have succeeded at dislodging power (at least on occasion) this may have happened for different reasons than those suggested by the "opinion model" implied by studies of coherent elite positions.

The lifestyle positions *could* be tamed into the categories presented by attitude studies; we might take lifestyle proponents at their word when they say they are trying to change opinions or are modeling a preferred future. I am claiming that such explanations are just too problematic to cohere; something else just must be at work here. As an exercise in forming public (and elite) opinion, the lifestyle gestures are incomplete and insufficient. The notion that one could build a society by "acting peacefully" is naive and limited. The whole lifestyle position presents itself in a fatalistic, simplistic, almost farcical way, inviting exclusion from serious study. We can almost hear Pat Buchanan screeching, "Hug the Soviets? Come on!"

So, if that approach "works" — if it influences events, if it is persuasive to serious citizens, as it seems to be — it might be working in some different way. What might that "different way" be? There could, of course, be many different answers; that is the implication of the loss of a social that could firmly ground our interpretations of political acts. Perhaps the intentional simplicity — a purposefully dumb stance — expresses the exasperation of citizens at the brink of destruction, but also at the end of their energies, after prevailing categories collapsed. Or maybe I have only emphasized the slogan, mistaking it for the analysis; some proponents of the position might point to sophisticated, in-depth analysis, situated just behind the slogan. And some activists would contest my abrupt dismissal of modeling and opinion molding; Mom's recycling didn't mean much on its own, of course, but it is how we learned about politics and the practices of everyday life.

The point of the demise of the social is that every interpretation does indeed hold a certain credibility, or, to put it more dramatically, none of the interpretations is grounded in any serious way. Strategies of explanation constitute their own audience, by means of mechanisms that contemporary literary criticism introduces. This is a momentous development for all political analysis, as I will discuss in the next chapter. For now, I can just forge ahead, opening yet another interpretation and using it to trace the rhetorical devices at work in this one, narrow part of the political landscape. Doing this kind of interpretation then has inevitable political consequence; every successful reading further undermines the usual discourses of foreign policy, bringing legitimacy issues to the fore, perhaps setting off a clash of discourses that establishes new (if temporary and shifting) political ground.

The interpretation—the "spin," to use the Reagan-era term—I want to consider goes like this. Not presuming to enter into the realm of force-counterforce (and all the other economies of force surrounding military and nuclear matters), the "lifestyle" argument *simply intervenes.* This intervention produces consequences that are more ironic than representational, more disruptive than analytic. The lifestyle position works by rubbing against a nuclearist discourse that has tried hard to exclude challenges to its logic. In its partial, deconstructive mode, that opposition has worked, putting its own "dumbness," its forced inarticulateness, against the forced coherence of foreign policy discourse. Arrayed against a thoroughly coded way of speaking, the opposition stripped its own utterances down to a naked minimum—not escaping code (how could anyone presume that?), but forcing the dominant discourse to handle the weight of the codes and substitutions all by itself.

My reading works, then, on language-and-politics turf captured by Foucault. I am postulating a specific kind of intervention—one that politicizes by noting how language works, without forfeiting the next political response. Foucault claims this odd and important double move with a distinctive two-part challenge to power. Starting with the crisis of representation and character of language that sets that crisis off—shifting and turning away from either the self who uses it or the phenomenon it tries to capture—Foucault moved on to a description of rules and the ways those rules constitute a generally unrecognized realm of power in contemporary society.[25] The two moves resonate, one exacerbating the other until legitimacy is drawn into the whirl of contested territory. Foucault's conception of language is what funds the possibility of political response, making it possible that such response is neither an arbitrary imposition, as has been charged, nor a promise of meaning and representation that cannot be fulfilled. Instead, the political response finds its form exactly at the point where old models of language break down.

In Foucault's analysis, language holds a different position than we have long expected it to hold. Its mechanisms turn out to be just as reversible as any political interpretation (which is, of course, where that political reversibility begins).

> Language is . . . freed from all of the old myths by which our awareness
> of words, discourse, and literature has been shaped. For a long time it
> was thought that language had mastery over time, that it acted both as
> the future bond of the promise and as memory and narrative; it was

thought to be prophecy and history; it was also thought that in its sovereignty it could bring to light the eternal and visible body of truth; it was thought that its essence resided in the form of words or in the breath that made them vibrate. In fact, it is only a formless rumbling, a streaming; its power resides in its dissimulation.[26]

Rather than fulfilling the role of the master (bonding future promise to past memory, illuminating truth through words), language actually threatens that role. Dissimulation, waiting, and forgetting are more central to language than illuminating (arriving at something that can be illuminated). Its failures forgotten in the continuing stream (or discounted while we wait), language works by seeming to be what it cannot actually be. While it seems to promise clarity about a future it will encounter, language still never gets there; "it is depthless forgetting and the transparent emptiness of waiting."

> Language, in its attentive and forgetful being, with its power of dissimulation that effaces every determinate meaning and even the existence of the speaker, in the gray neutrality that constitutes the essential hiding place of all being and thereby frees the space of the image — is neither truth nor time, neither eternity nor man; it is instead the always undone form of the outside.[27]

If language is "neither eternity nor man," and is continually undone, it forms an oscillation, giving only momentary contact with either self or phenomenon. It is a continually "undone" form.

Anyone who relies on such a form, which is to say anyone who speaks or writes, will find their positions brought into question, into the vertiginous gap Derrida calls the *aporia*. After all, every political claim originates in language. Furthermore, while the earlier, mistaken view of language that Foucault criticizes had granted a special role to the speaker — the author — this new position undermines the self. Neither originator of meaning nor dupe of a determined, exterior script, the self is constituted in a special way. By settling into "neither truth nor time," language "places the origin in contact with death, or rather brings them both to light in the flash of their infinite oscillation — a momentary contact in a boundless space."[28]

Rather than solidifying our relations with events and their sequences, language inserts oscillation, dissembly, memory, and waiting. The attempt to solidify foreign policy positions thus runs counter to what occurs within those positions, at their most fundamental levels. But instead of making politics impossible, our recognition of language's shifting, contradictory role enables responses, even if those responses will differ from those enabled by the Enlightenment.

At this point "the self" becomes important to my reading of the lifestyle position. The oscillation and contradiction rearrange the role of the speaker, that most central role in any politics that could emerge from the Enlightenment. Even "the simple assertion 'I speak' " contains the speaker's (and the statement's) "threatening promise of its own disappearance, its future appearance."[29] Losing the status we had assumed the authorial self had, an oscillation begins in politics as well as language. The undoing undermines some political speech; the exterior, moral, critical distance promised by Habermas cannot hold. Neither can our commitment to re-enact the world we prefer through alterations of grammar and syntax. But the same undermining has already (long ago) happened to institutional structures and authorities who have imposed political forms. Those structures cannot be justified by the web of relationship to (the words) nature and necessity. The "future appearance" would be uncertain, to be sure, but it would also threaten all political forms that drew their justification in this manner. Of course, this encompasses virtually every contemporary, institutional, political form of any importance.

Foucault summarized these effects of language and power by referring to practices that constitute a new variant of the "self." We have been looking for power to operate in some overt way—open in its operation and universal in its application—but power has assumed, instead, a form separate from agents and conscious action when it operates, as it now does, through language and habit (an ensemble he called discursive practices). If "selves" are constituted, if the habits of social life have a logic that precedes the pursuit of moral order, then the status of personal action has been severely undermined.[30] Feminism marked the first form that began its politics by acknowledging that constitutive basis and pressing "personal" claims to politicize relations, more than to solve them. Following from that lead, a host of political claims, which diverge radically from the issues and ideology of the past, have emerged over the last two decades. Utterly dispensing with traditional issue positions, an entire generation of political activists has now taken a different approach.

I do not suggest that the lifestyle position has been somehow covertly poststructuralist all along. Indeed, that position seems to have arrived at its ironic, interventionist, and still oppositional position from entirely diverse directions. But the effect is clear. Steadfastly ironizing all discussions of force into peace, exuberantly refusing the (previously obvious) complexities of foreign policy and arms control, dislodging universalisms such as rationalism into personal practice, localizing the critique of what everyone thought were manifestations of global forces—all of these are deconstructive tactics. In their wake, each pronouncement—whether from a president or a spokesman of the State or Defense department— necessarily will be simply another spectacle.[31] News accounts more and more obviously come to construct the problems they analyze, creating predictable subplots of threat and reassurance, tension and release. "Observers and what they observe construct one another."[32] The reference to a coherent, unproblematic social rationalism cannot hold.

Put another way, we might now begin to understand that President Reagan's famous slips and spectacular flights of fancy were neither idiosyncratic nor entirely "his fault." Rather, they marked an inevitable development, following the demise of the social and the deconstruction of rationalisms; even foreign-policy politics would begin to lose its grasp on a convincing discourse. Irony holds what used to be the center. Even the president of the United States must stand nakedly asserting power on the basis of spectacle, not fact, whether social or technological. His statements—the performative utterances of entirely real power—would have to take the same form as the antinuclearist slogans I have been making jokes about. Symbols and codes begin to break free from their anchors:

> Reagan had remarked extemporaneously at a press conference that the only parts of SALT that were being observed were those that had "to do with the monitoring" of weapons. As so often when Reagan was discussing arms control without the benefit of a carefully prepared script, he had his facts wrong. Pressed on why his . . . proposal bore down so hard on land-based, as opposed to submarine-launched, ballistic missiles, he explained that SLBMs were less threatening because, like bombers, they could be recalled after they were launched. This came as news to the U.S. Navy. . . .
>
> Commenting on the strategic triad, Reagan noted that the most destabilizing weapons were land-based missiles because they were "the biggest and most accurate." So far so good. But he continued: "Also, land-based missiles have nuclear warheads, while bombers and submarines don't." Even as he said these words, his voice dropped and wavered, as though he had forgotten his lines and knew there was something not quite right about his attempt to improvise. . . .
>
> It fell to [Admiral Bobby Inman, deputy director of the CIA,] to tutor the President on the most elementary aspects of the subject at hand. Reagan asked, "Isn't the SS-19 their biggest missile?"
>
> "No," replied Inman, "That's the SS-18."
>
> "So," said Reagan, "they've switched the numbers on their missiles in order to confuse us!"
>
> Inman smiled. "No, it's we who assign those numbers to their weapons systems in the sequence we observe them, Mr. President."[33]

## Irony and Politics

If we now reassemble the positions I have been discussing (on language, the social, and the self), some distinct political possibilities emerge. One way to summarize these positions would be to focus on irony, the pattern of speech closest to this notion of language.[34] Displaying an unwilling-

ness to objectify discourse—uneasy about being pinned down—the ironist shifts, or oscillates. Critique and program become untenable. And, as
with any aesthetic pose, the possible existence of some audience is central
to the ironic discourse, even if the performative, power-building potential
recedes.

Failing to understand how their interventions worked, the proponents
of "personal is political" failed to sustain the ironic base from which they
had emerged, quickly retreating to liberalism when discomfort about the
new condition of the social realm and the self became acute. To focus on
antinuclearism, there was an unmistakable tone of irony in the movement's American renaissance during the middle 1970s. Fresh from other
struggles, radicals pursued a studied "cleanliness" (against leaks) that actually aimed at the heart of a very dirty matter: corporate and military
practices. The simple claim—let the sun shine—obviously stood for (signified) a controversial and dangerous claim; balance should be positive
and harmonious, in contrast with the studied discord of deterrence.

These practices of irony have a familiarity in American culture, most
publicly in the art world. When popular art learned to use them, teaching
the usefulness of ironic passion to the politicos, it made the aesthetic into
a political act. It is in popular arts—at their margins—that the ironic has
survived, even while the political movements seem to have lost their lessons. Listen to Elvis Costello sing "(What's So Funny 'Bout) Peace, Love
and Understanding," for example.[35] Anguish and uplift—a truly incongruous pair—combine, seeming to complement one another. The singer
could be satirizing a too-serious hippy, or he could be issuing a complaint. Asking for harmony in a song that has no voices to accompany his
own, Elvis Costello (whose name also combines the hyperserious buffoon
and the buffoon) serves as example of expression without alienation, of a
potentially angry politics that has somehow survived the demise of any
steadied narrative.

Acknowledging the ironical contradictions at the heart of any objective world, antinuclearists went giddy with code and symbol, purposely
arraying each code against itself. They took a position that only seemed
to be a coherent, oppositional unit. The irony of a "clean reaction" to the
nuke—a gentle and attractive lifestyle—recalled Lockean citizenship, but
did so in a context of absurd, worldwide aggression and deterrence that
mocked the village model of a classical liberal politics. A context was appropriated that would render every attempt at sincerity absurd, a reso-
nating ensemble of contradictions so thoroughly designed that it could
not have been naive or mistaken. The modesty could only be ironic.

The self-conscious joy at taking this perverse position has no doubt
dimmed in antinuclear communities. The "failure" of the approach could
be variously described. One explanation is that "politics moved away"

from the antinuke position, rendering moot the sort of interpretation I am offering. Ferguson's description of the whole system moving rightward, which I discussed earlier, invokes this type of explanatory device, one firmly lodged in a larger discursive construction that allows (then requires) the ranking of positions within a fairly stable left-right continuum, granting an aura of reality to those positions that is undeserved, even from within conventional "policy approaches."[36]

It could be that the ironic simply went underground—unrecognized but still present. John Seery's reading of an antinuclear demonstration leaves open just such a possibility. Closely reading the actions of demonstrators, he finds striking examples; a handicapped woman puts herself (in her wheelchair) in the way of the nuclear enterprise, and is arrested. As Seery explains, that action has to be ironic, and is coherent, as well as potentially effective.[37] Seery is unable, however, to interview the protester; it is his reading that preserves the ironic in this case. He does not consider the other possibilities—that antinuclear politics could be naive, or that it could (egotistically) take its own warnings as instrumental, rather than symbolic. Surely, this is because his reading of the demonstrator's actions grants them the dignity of an ironic interpretation. But, at the same time, Seery knows that the ironic possibility is a fragile one; "irony is difficult to interpret or to sustain. As a strategy of nonviolent political resistance, irony carries with it no guarantees; it can fail. The Athenian court misread Socrates' irony. The English largely misunderstood Jonathan Swift."[38] Seery hopes that the protesters he calls "the modern-day gadflies of the nuclear age" will be better received, but the fragility of a politics of irony is visible, even within his analysis.

Joseph Gusfield offers another explanation for the demise of irony, reading its workings at a more personal level. At the end of his book *The Culture of Public Problems*, in which he had dramatically reread the policy discourse surrounding drinking and driving, Gusfield recounts an incident where his analysis met a strong response from a practicing physician. The doctor argued that Gusfield's analysis ignored the pain generated by alcohol-related accidents; Gusfield "justified his earmuffs" by describing his position as a "sociological irony."[39] In one sense, a certain distance from events is necessary for the analysis of social events, he argues. This may well create a tension with other impulses—nobody wants to ignore the suffering of a victim injured for the least rational of reasons. But the ironist requires a certain distance in order to make the familiar seem suddenly strange; the ironist doesn't (and couldn't) work in the clinic, alongside the physician.

Transposed to nuclearism, Gusfield's argument takes on some odd twists. On one hand, we begin to glimpse one reason for the prominence of the "lifestyle" approach; faced with a crisis that is associated with mass (if future, imagined) suffering, people would prefer to respond in a modest, caring mood, perhaps reacting as if we were already at the bedside of the victims. That impulse drives much of the antinuclear position; in books, speeches, and films, we are asked to imagine (unimaginable) carnage, and to gauge our reactions accordingly. The ironist cannot participate in such exercises, as Gusfield explains; "One cannot engage in irony without assuming a distance and detachment from those being described."[40]

Still, Gusfield is careful to explain the political dynamic underlying sociological irony. The inquiry he undertakes necessarily opens political ground, when it examines the symbolic nature of authority and offers other interpretations. "To find alternative ways of seeing phenomena is to imagine that things can be otherwise. . . . This cannot but be a diminution of the legitimacy which authority gains from a belief in its facticity."[41] This is in particular contrast with the view of irony expressed by Richard Rorty, for whom a "liberal irony" leads to justifications of a political quietism.[42] This is a dramatic difference of opinion; two respected scholars, one a social scientist and the other a philosopher closely attuned to the social sciences, examine and adopt irony as characteristic of their position, then come to very different conclusions about its implications.

The ironic offers no standard by which we might declare Gusfield correct and Rorty false. Gusfield uses distance to generate politicized explanations; Rorty generates distance in order to establish a rather complete distance from politics. Since both positions are ironic, they are both skeptical about any single test that would reconcile these two possibilities. There is one more thing to note about these two well-known positions on irony. Both Gusfield and Rorty are both speaking from the standpoint of the observer, asking questions about how one understands events from a distance. Both edge their discussions toward politics by taking themselves as a kind of model; neither falls into the old positivist trap of imagining they can stand fully apart from the events they study. Still, their irony is a feature they self-consciously adopt, urging others to consider it.

There are several points to consider before we apply this notion of ironic choice to nuclear politics. In Gusfield's case, the ironic position is driven by his wish to "emphasize the existence of a nonutilitarian, symbolic element" in public interactions.[43] As I have been arguing, the nuke has always had this symbolic element, and it is much harder to avoid that aspect than would be the case in other policy areas, such as the drinking and driving example Gusfield is considering. This is not to say that this has been an easy adjustment for nuclear-age citizens to make; this symbol-laden question meets a culture that privileges the utilitarian character of public interactions. For example, we encounter such symbolic interactions as the nuclear balance of terror with a long-established, well-learned understanding that the state has a monopoly on coercion. Both private acts and expectations are problematized by the shift toward the symbolic. Nonetheless, this irony is not really an "observer's choice." We nuclear-age citizens all find ourselves embroiled in an ironic setting. This is no longer an observer's choice, but a feature of the situation, which we may try to avoid (being good, apolitical Americans) but cannot.

This take on the ironic practices of nuclearism replicates a condition Baudrillard calls *external irony*. This irony is no longer merely a feature of the response offered by protesters and analyzed by observers, but has already consolidated at a more general level. From Baudrillard's position, the entire system of difference tends to reverse and collapse. Trying to assume an external position, one ends up joining an endless chain of duplications. "In trying to be oneself, to cultivate difference and originality, one ends up resembling everyone."[44] As that external position collapsed, the antinuclear movement became increasingly liberal or, perhaps, neoliberal.[45] Made with surprising ease, the subtle shift from irony to values precisely followed a trajectory Baudrillard repeatedly identifies:

> We all remain incredibly naive: we always look for a good usage of the image, that is to say a moral, meaningful, pedagogic or informational usage, without seeing that the image in a sense revolts against this good usage, that it is the conductor neither of meaning nor good intentions, but on the contrary of an implosion, a denegation of meaning (of events, history, memory, etc.).[46]

But the image revolts not only against these external critiques, but also against the established system. This is what makes possible a new ironism, a politics of intervention that could yet grow from antinuclearism's collapsed critique.

> Today we have a form of irony which is objective. Irony can no longer ... be simply the subjective irony of the philosopher. It can no longer be exercised as if from the outside of things. Instead, it is the objective irony which arises from within things themselves — it is an irony which belongs to the system, and it arises from the system itself because the system is constantly functioning against itself.[47]

Obviously, this "functioning against itself" is not simply Marxist or dialectical. It is, instead, the dysfunction of language and politics; without a social realm that serves as that which is signified, mainstream positions (as well as their critics) begin to shift and float, as Derrida has persistently noted. Not susceptible to liberal or Marxist response, the nuclear culture turns out to have long been vulnerable to itself.

This "external irony" was often misdiagnosed as existentialist absurdity by the American counterculture of recent decades. Playing roles, not always serious about one's own persona, this part of their politics was inadvertently ironist or, rather, was ironist to a point (that point being the now caricatured search for alternative metaphysical ground that marks every enterprise called "New Age"). Taking the frivolous seriously, refusing to do the same for the potentially fatal, renegotiating boundaries through continual reversals (of roles, expectations, perspectives), these amateur ironists briefly experimented with the future of oppositional politics in a postmodern world.

As in every other political arena, the critique of poststructuralism in foreign-policy politics repeats one theme.[48] There must be an external viewpoint from which to issue commentary, this critique always claims, or else there would be no reason to resist: no concretized goal, no preferred future. Beneath that critique lies a confidence in certain versions of "the real." But, no matter how strenuous the criticisms, every attempt to track this "reality" leads quite directly to the political constructs Foucault describes as the constant form of political speech. And we can answer the criticisms even more directly. Simply put, the partial interventions (or, as Spivak calls them, interruptions[49]) performed by various literary criticism approaches have already been effective, repeatedly and in diverse circumstances. The issue is only how to lodge such interventions in a discourse that can continue—once it understands itself as opposed to "realistic solutions"—to engage in a struggle against disjointed but effective power.

Late in his career, Foucault defined the context for his politics as a personal ethic, assembled on aesthetic grounds, in what Bernauer suggests amounts to "a new effort by Foucault to integrate his historical researches with the issue of the subject who must act, who is incited by that research to transform the real."[50] Instead of focusing on liberation—an ambivalent condition, which may be susceptible to the flaws of ideology—Foucault instead insisted on "the practices of freedom," an ethic.[51] In the last works published before his death, Foucault found traces of such an ethic in ancient Greece, where he found "an entire ethics that turned about the care for the self," giving "ancient ethics its particular form."[52]

Foucault's last challenge, then, was to found such an ethic on an intervention into the production of truth, not on the use of truth (or, more precisely, a truth-producing discourse) to produce policies and positions. Indeed, it is in exactly this context that Foucault makes one of his only comments on what he calls "the ecology movement," a political stance that is a part of the lifestyle position I have been discussing:

> There has been an ecology movement—which is furthermore very ancient and is not only a twentieth century phenomenon—which has often been, in one sense, in [a] hostile relationship with science or at

least with a technology guaranteed in terms of truth. But in fact, ecology also spoke a language of truth. It was in the name of knowledge concerning nature, the equilibrium of the processes of living things, and so forth, that one could level the criticism. We escaped then a domination of truth, not by playing a game that was a complete stranger to the game of truth, but in playing it otherwise or in playing another game, another set, other trumps in the game of truth.[53]

The point for the lifestyle proponents may be to present their position as enduring irony, unlikely to be domesticated, resolutely irresolute. The model here is the artist: back and forth, eddying from criticism to expression before starting again, differently this time.[54] No certainty is available; the language-and-politics position has accurately observed the end of that possibility. But the possibility for interventions against power is hardly made unfeasible by this lack of certainty.

Foucault concluded his brief discussion of ecology politics with a calibration of just such an opening:

> We could criticize politics—beginning for example with the effects of the state of domination . . . —but we could only do this by playing a certain game of truth, showing what were the effects, showing that there were other rational possibilities, teaching people what they ignore about their own situation, on their conditions of work, on their exploitation.[55]

The progressive hope can be discarded—must be, indeed—but the notion that politics itself vanishes at the same point is a vain critique, one that conflates utopia or hope with the struggles against power. It is already clear, I have argued, that the struggles continue quite well under the flag of (unflagging) irony.

The demarcations between such an ironic position and a more sincere one will always be a matter of interpretation and negotiation. If one starts with the assumption that politics must be oriented from some external, confident position, anyone who denied that assumption is then open to the charge that he or she has engaged in purely personal indulgences and extravagances, with no link to institutional change. It is a subtext of my argument to demonstrate how far this assumption is from being a necessity.

Once more: the abandonment of any metaphysical solidarity in no way requires the adoption of private, solipsistic signaling into the void as a model for politics. But the dispute resolutely remains lodged *in between* truth-producing discourses. Any case for (or even against) the ironical, interventionist position will always be speculative and rhetorical. It is, at a minimum, important to dislodge the presumption that the lifestyle position represents the only (or even necessarily the first) response possible in the face of nuclearism, before I move, in the next chapter, to expand discussion of alternatives.

In this chapter, I have tried to reread one of the most self-consciously direct and literal political moods in the contemporary era. If I can make even such performances as a "Think Peace" bumper sticker seem *political* in character—capable of multiple interpretations, ironic, and aimed at the workings of power—there may yet be reason to hope for an interruption in the nuclearist discourse.

# 6

# Power/Cheekiness

Looking back on the first half century of the nuclear age, we can begin to plot the questions and answers that have had an absorbing effect in politics. If all of the extraordinary events of 1989–90 have been surprising, that may suggest we have missed some crucial aspects of this age.

These dramatic political events have not yet produced an analysis that would match the spectacle of the events themselves. Most analyses emphasize "popular will," "freedom," and the triumph of capitalism over communism, but each of these points had been undermined long before 1989 gave them a last breath. Even the briefest interpretation of these events deconstructs such ideological readings; for example, socialists frequently were on the popular side, as in South Africa, China, and most of Eastern Europe. Narrow, legalistic readings of freedom were put into question by the broad, cultural (and, sometimes, ethnic) character of many of the outbursts. And the "popular will" asserted itself so suddenly—especially in some instances in Eastern Europe—that such a will becomes a suspect explanatory device.

As I have been suggesting throughout this book, some of the most obvious, "commonsense" political positions have turned out to be less useful than we thought they were. Specifically, the claim that a straightforward concern for survival might underwrite a new antinuclearism encounters a challenge. The humanism this concern relies upon tends to break up, with rationalism giving way to irony and cynicism. In this chapter, I will reexamine this political terrain, noting especially how Jean

Baudrillard and Peter Sloterdijk can assist an understanding of this new and odd situation.

## Humanism

Overall, the Western press and intellectuals explained these events in broadly humanistic ways, ignoring in the process some longstanding warnings. Foucault offers one of the most prominent (surely not the only) commentary on humanism as an explanatory device:

> This idea of man has become normative, self-evident, and is supposed to be universal. Humanism may not be universal but may be quite relative to a certain situation. What we call humanism has been used by Marxists, liberals, Nazis, Catholics. This does not mean that we have to get rid of what we call human rights or freedom, but that we can't say that freedom or human rights has to be limited at certain frontiers. . . .
>
> What I am afraid of about humanism is that it presents a certain form of our ethics as a universal model for any kind of freedom. I think that there are more secrets, more possible freedoms, and more inventions in our future than we can imagine in humanism as it is dogmatically represented on every side of the political rainbow.[1]

A full review of humanism is far beyond the scope of my present project, but some correctives are suggested as an offshoot of the nuclear age. Surely, Foucault inaugurated a deconstruction of any assumed "depth humanism" that has broadened since his death; the use of elemental human impulses to explain political events has been problematized, no doubt irrevocably. Reliance on "truth-telling" as a political response also comes into question. This is not necessarily a matter of events being complex or deeply deceptive; after all, as Baudrillard suggests, our era can be described as one in which all is on the surface, without depth to plumb. In an era when a surface contradiction—not implying a determining structure, but operating just as it looks—dominates, we have responded by a peculiar compulsion for the truth, and with a naive confidence in authenticity.

A conclusion emerges from several of the discussions in this book, including Kateb's critique of the "survival" position (chapter 1), the issue of the warrior archetype (chapter 2), and the question of deterrence discourse that runs throughout the book. This conclusion, in summary, is that humanistic approaches have injured our ability to respond to this new and peculiar world. And they have injured our cognitive abilities, as well. Changes in the nature of power have been visible to theorists for a long time; now, these changes have erupted throughout the political world, but have yet to set off a new analysis—one which might account for the new power lodged in *the event*.

As I discussed in an earlier chapter, Reagan's triumphs should have served as an introduction to a new moment of power, one in which an intense presence of the image produced a presidency Diane Rubenstein has called "the most perfect exemplar of Baudrillard's . . . simulation." The Reagan era matches each feature of simulacrum in "the neutralization of the signified by the code, the radical displacement of the referent by the model and the priority of reproduction over production."[2] While the Reagan age continued, critics from the left emphasized the simulation as its weakness, as a stasis funded by lies; frustrated by the end of activity *as they knew it*, the left assumed that nothing was happening.

Even in the early 1980s, however, Baudrillard had identified the tragedy of the left's waiting for events to propel themselves along the course the left had charted—that is, the world's propensity for a hyperreality that would soon erupt in unmistakable ways: "All we may expect of time is its reversibility. Speed and acceleration are merely the dream of making time reversible. You hope that by speeding up time, it will start to whirl like a fluid. . . . As linear time and history have retreated, we have been left with the ephemerality of networks and fashion, which is unbearable."[3] This is "unbearable" in terms of our predisposition to await and foment the unfolding of events; that is not the same thing as stasis or gridlock (or, on the other hand, the resolute fix of interests and conspiracy). This new order of events, of the world, may have been obscure when Baudrillard first wrote of it, but it would become unmistakable in the late 1980s.

Gorbachev's trajectory through recent years seems a fulfillment of Baudrillard's *Seduction*, not Jefferson's *Declaration of Independence* or Marx's *Capital*. The theme was challenge and seduction; offering the West what it asked for but could not bear to have (arms control first, then the change in the status of Eastern Europe), turning liabilities into triumphs (in the repeated Communist Party "catastrophes," many of which turned to his advantage, at least until the breakdowns of late 1990). China, Eastern Europe, and South Africa in 1989–90, the Middle East in 1990. The rules have clearly changed, and we are left to try to figure out what happened.

Faced with this list of important absurdities, twists, and reversals, we might look to political science for help; surely, that discipline would have developed some interpretive strategy for untangling the problems power brings forth. Such a project has emerged, but only at the margins of the social sciences, as in the works I have cited in this study. Political science has surely been attuned to the machinations of power, and that has been the fallback position of political analysts trying to talk about Gorbachev.

But political science has always relied on an underlying order—a signified—that is represented by power politics. Whether Marxist or democratic-pluralist, there had to be a base. It is only the rare political scientist who has noticed how completely actual politics can swirl around that presumption of interest.

At the same time, power is Foucault's theme in his response to humanism. The liberal or humanist wants to vault over power, superseding it by an appeal to a humanism that can establish critical distance. Foucault's cautionary note emphasizes that the idealist form chosen by the humanists allows power to continue on, in new and even more potent forms. The key to reestablishing a politics after that series of events has happened is in the "constitution of subjects," a topic long ago identified by Murray Edelman, a crucial figure in the emergence of postmodern analysis within political science:

> If politics is concerned with who gets what, or with the authoritative allocation of values, one may be pardoned for wondering why it need involve so much talk. An individual or group can most directly get what it wants by taking it or by force and can get nothing directly by talk. . . . The employment of language to sanctify action is exactly what makes politics different from other methods of allocating values. Through language a group can not only achieve an immediate result but also win the acquiescence of those whose lasting support is needed. More than that, it is the talk and the response to it that [measure] political potency, not the amount of force that is exerted. Force signals weakness in politics. . . . Talk, on the other hand, involves a competitive exchange of symbols, referential and evocative, through which values are shared and assigned and coexistence attained.[4]

The 1967 book in which that passage appears, *The Symbolic Uses of Politics*, was a major success and has educated a generation of students. But the generation of professors who assigned it did not move immediately to a recognition of politics as symbolic and discursive. Instead, they tied the symbolic to the manipulations of visible and well-known power, when they bothered at all with the realm of language. Edelman's work was read as an explanation of how dominant power had forestalled a march of events much better understood by structuralisms of interest and personality than by any new regime of rhetoric, narrative, and discourse: what Edelman calls "so much talk."

The difficulties of political science may not be of much interest in themselves, but they do set a direction for further inquiry. The much-maligned renewal of interest in French social thought surely implies a bypassing of American social science, a way of circumventing it. Of course, French thought encounters resistances of its own, as well. Frequently, however, figures such as Foucault, Derrida, and Baudrillard have taken a different tactical stance with respect to the challenge of dominant modes of thought, frequently practicing the outrageous as an affront to a steadfast orthodoxy that yields none too easily to arguments it would recognize.

## Rationalism and "Politics with Attitude"

I have been suggesting, throughout this book, that Baudrillard et al. might provide a useful starting point for political analysis and movement. At first glance, Baudrillard would seem to be the last outpost along the path of the demise of politics; he says as much, repeatedly. Still, it seems extraordinary to me that American readers have so often assumed that this put him in the camp of the "end-of-politics" theorists with whom he obviously had nothing in common. There seems to be an aesthetic of social thought in this country; rationalistic stylings imply seriousness, understatements suggest breadth, utopian goals and values stand for realism, and so on.

The aesthetic of political rationalism—more specifically, a rhetoric—assumes a certain directness of purpose, perhaps antidialectical in origin, but now more general than merely an antimarxist tactic. There is some reason, then, to choose Baudrillard as a keynote for a political response; much less likely a starting place than, say, political science, he ends up being a perfectly appropriate place to begin. It may not be so much paradox as irony that moves us toward Baudrillard—who has been so emphatic in claiming this end to politics—as a place to start a postmodern move on political space. Humanists conflate their tactical approach with all of politics; Baudrillard's response both criticizes and makes fun of their solemn voice-over.

These matters of style and rhetoric can, after all, be crucial. There is some evidence that Baudrillard has been wildly misread on this count. Calvin Thomas's important reinterpretation of Baudrillard's *Forget Foucault* makes just such a case. "For Baudrillard, Foucault's mistake is to allow his discourse to mirror flawlessly, and with its own panoptic perfection, a power that everywhere produces and reproduces the social as 'the visible, the all-too-visible, the more-visible-than-visible.' " Thus, what has been oddly read as a move to the right by Baudrillard turns out to be something altogether different: "Baudrillard wants to reveal to Foucault the postmodern abyss that opens up in his own discourse, to activate Foucault's will to disappear into that ironic space. This is Baudrillard's seduction of Foucault, his challenge to the death: against the power that is everywhere, the seduction that is elsewhere."[5]

"Elsewhere" is a difficult position to establish, when one is trying to establish difference from a position that has successfully claimed hegemony for so long, repelling so many previous assaults. To accomplish that challenge, Baudrillard adopted a rhetoric of indirection that intends to confound. Accordingly, it may be useful to pair Baudrillard with another theorist whose take on rationalism is more direct. For Peter Sloterdijk, the challenge has always been with us, as long as philosophy has made rationalistic claims:

> For the philosopher, the human being who exemplifies the love of truth and *conscious* living, life and doctrine must be in harmony. The core of every doctrine is what its followers embody of it. This can be misunderstood in an idealistic way as if it were philosophy's innermost aim to get people to chase after unattainable ideals. But if philosophers are called on to live what they say, their task in a critical sense is much more: to say what they live.

Diogenes—whose appearance on the scene "marks the most dramatic moment in the process of truth" in early Western philosophy—confronts the impossibility of Plato's rationalism. The significance of cheekiness, which Sloterdijk portrays as very broad, arises from this confrontation:

> Since philosophy can only hypocritically live out what it says, it takes cheek to say what is lived. In a culture in which hardened idealisms make lies into a form of living, the process of truth depends on whether people can be found who are aggressive and free ("shameless") enough to speak the truth. Those who rule lose their real self-confidence to the fools, clowns, and kynics.[6]

## The Ironic, the Abyss

With both Baudrillard and Sloterdijk in place on the tricky topic of rationalism in politics, we can begin to account for the serious criticisms of this approach. If Baudrillard, Sloterdijk, and others went giddy finally to escape the rationalists, they certainly found critics willing to take them on.[7] What has been less widely discussed is that Baudrillard's style of writing is fully a part of his work. Directing one's theory toward the goal or end of developments is obviously a rhetorical position of long standing in Western social thought; reversing that tradition—or even bringing attention to it—is no simple matter. Baudrillard frequently makes it clear that this posture toward ends is now in play. In *America*, he makes the point repeatedly, as when discussing the automobile: "The point is not to write the sociology or psychology of the car, the point is to drive."[8] As he makes clear elsewhere, there is no doubt that this "drive" is a general issue of approach: "Just as the world drives to a delirious state of things, we must drive (slowly) to a delirious point of view."[9]

In short, Baudrillard is working at the extremes of each phenomenon in an era when, he argues, every event reaches toward extremes. Still, this is more than a matter of hyperactivity; in his remarkable discussion of the California desert, Baudrillard makes the point that the late modern era epitomized by America has an identifiable attitude toward its goals: "[The desert] is a hyperreality because *it is a utopia which has behaved from the very beginning as though it were already achieved*. Everything here is real and pragmatic, and yet it is all the stuff of dreams too."[10] As a recent discussion suggests, Baudrillard has adopted this position toward extremity — writing as if the horizon is a whirl of disconnected signs and assumptions, not a rational expectation nor a condition created by our practices of resistance:

> I don't feel — when I am writing — I don't feel it as a political act. I feel it maybe as a symbolic act. Maybe it is a fatal strategy itself — a theoretical fatal strategy — and maybe it has some symbolic effects, to accelerate. Not to resist, I'm not trying to resist, surely. But, I don't know, maybe somewhere I have some, not hope, but some opening to the void. Maybe there will appear, out of the absence of representations, some new events, even more fascinating than past events. But we cannot create them out of our own will or representations. That is sure. Then, anyway, it will be a suspension of the probability of events. There already have appeared some new events. Terrorism is a new event [, as are] AIDS, viruses, or Wall Street crash. These are new "objectile" events. Our time is very poor — at least in France — as to subjective political traditional themes. But our time is rich — over-rich — in metonymic events.[11]

Clearly, this posture toward extremity informs Baudrillard's statements on the politics of the nuclear age. The accumulation of events in 1989 and 1990 followed a path Baudrillard surely understood; Gorbachev exemplified an "openness to the void," and the onrush of events was the major player, as the geopolitical shape of Eastern Europe altered. In terms of predictive quality, Baudrillard's precedent may have been unlikely, but it surely outstripped the more sober visions of, say, international relations scholars, journalists, and historians in the West.

Baudrillard's work continually reenacts a poststructuralist movement beyond constraining structures and depth. The crisis of representation surely deprives events of (economic, psychoanalytical, or religious) depth; its confirmation would be found in a reversibility of explanations. Two contradictory explanations might coexist, remaining untested by depth analyses that had lost their authority. A kind of lightness attains, this time a lightness of meaning.

This idea of reversibility — the tendency of postmodern events to elude some essential meaning in a dizzying, unconstrained sequence of change — is more than lit crit jargon. As Arthur Kroker has noted, there is even a history of reversibility waiting at the core of liberal institutions. U.S. political life has always had an unexpected reversibility at the center of its notion of power, long before Foucault wrote of doubles and Derrida slipped away from the scene of essentialist politics.[12] The American founders took thorough precautions to insure that one set of governors could be replaced by another. The Enlightenment's early prediction of public opinion—expressed through the crude and slow mechanism of the vote—would be the vehicle for that almost unprecedented reversing of power.

But when vehicles or solutions "go bad," wild consequences occur. Reversibility spins as far out of control as invisibility did in those old "invisible man" movies. Just so. The man is invisible; no version of the citizen is reliable enough to cover the diversity of phenomena required in the wake of Walter Lippmann's apology for that citizenry's demise. John Dewey's sincere response to that apology is a submission to a sucker's game; it cannot be won, just as the invisible man can't make it back to the world of the visible, where it once seemed easy to be. What is left after invisibility and suckers' games is public opinion, an aggregate and a trick, but—as Baudrillard shows—a sham, too. But the absence of a reliable public hardly deters reversibility at all, and that is the core insight of postmodernism in political analysis.

Indeed, the disappearance of "the public" augments reversibility, driving it to speedier reversions and more imaginative results. Change accelerates, as Chinese, Russians, Filipinos, Iraqis, Kuwaitis, and even Panamanians now know. The rootlessness of that change only makes it move faster, featuring developments whose great drama derives, in part, from the fact that they cannot be easily explained. If Americans do not yet know about that pace, it is only because they have yet to imagine what the others have begun to fathom; there will be an American Gorbachev, even if her visible birthmark will be placed differently.

If there is a problem with Baudrillard's "fatal strategy," then, it is not simply that he is lax with the details other writers take more seriously. There are good rhetorical reasons to pursue the sort of "wild social theory" Baudrillard has designed. But in the case of politics, it seems to me that Baudrillard misses something in his general statements. Baudrillard's repeated diagnosis of the "end of politics" requires further exploration. Surely his earliest works suggest that he is deploying his analysis against certain forms of Marxist class struggle. His extraordinary interpretation of Nixon's "silent majority" broadened his notion of the political to include the mechanisms of public opinion, and by implication the interest-group pluralism taken so seriously in American politics. As Baudrillard moved to consider contemporary politics—including, in *America*, the phenomenon of Reagan—his scope grew to include electoral politics. Beyond Marxism and liberalism, Baudrillard's treatment of politics becomes more difficult to trace. Opposed to theory, he never defines what this "politics" is that is disappearing.

Surely some old (modernist) versions of the political are indeed gone, for reasons Baudrillard describes persuasively. But the attitude toward knowledge and action that Baudrillard expresses is not entirely alien to politics; the political actor knows about objective irony, about the negotiability of the real, about the force of speed and visibility in contemporary society. This is what Foucault meant when he said he was an empiricist, a claim Baudrillard has echoed. The political forms that would enable these features while avoiding the problems now associated with representation are not at all obvious; my discussions of the freeze, for example, had to be abstracted and bracketed. But the stance toward the world Baudrillard takes does not abolish politics as broadly as he sometimes says it does.

The most important vanishing political element may well be the citizen-as-authority, as "author" who writes its interests and demands onto representative structures. To announce such a disappearance is surely to take a position that is anathema to liberal democratic theory, in which figuring out power is integrally a part of figuring out what to do about it, and then actively creating overt, reasoned responses that can be justified, announced, and formulated in terms of interest. But even this position, in itself, does not account for the general disappearance of the political. As the aesthetic turn in political and social commentary has demonstrated, there are other political moves than the impulse to write demands onto representative institutions. The illiberal or aesthetic choice has long been understood, within dominant forms of social thought, as dangerous and cantankerous.[13] This may be why postmodernism has been read as a flight from politics: perhaps better that than the dangerous politics it actually unleashes.

## Politics and Open Secrets

Perhaps the most obvious political consequence of depthlessness, reversibility, and acceleration is that events can indeed operate on the surface Baudrillard so often invokes. The surface is of interest to him for its openness; transformative structures don't intrude, visibility is unimpeded but unrevealing, and a particular disconnection is the prevailing mood. The surface—an antimetaphor, in Baudrillard's usage—becomes a basis for social and political analysis, often reversing commonplace criticisms. For example, in California, "culture itself is a desert . . . and culture has to be a desert so that everything can be equal and shine out in the same supernatural form."[14]

The "open secret" is not Baudrillard's term, but I think it captures the political status of this surface. In American political usage, to call something an "open secret" is to remind one's audience that this is something they already know, a significance they have already grasped, even though they have failed to act. Rhetorically, the open secret has served as a call to responsibility and action; no further information or argument is necessary. The open secret can also be evidence of a concealment (however failed); there must be some reason it can still be called a "secret," even though its secrecy has been lost or discarded. This reason could be a reference to depth; perhaps something remains an open secret because we cannot bear to deal with it, given certain elements of depth psychology. Alternatively, the secret could be a mark of corruption; the open secret is a reminder of unsavory choices made. In any case, this is the operative mood or attitude by which the real world of politics fulfills the role Baudrillard writes for it.

This dual character—both surface and depth—makes the open secret an exemplar of postmodern reversibility. Politically, the open secret would work just fine for antinukers such as Helen Caldicott, or even Robert Jay Lifton; theirs are theories of the truth we weak humans hesitate to face. But, just when guilt and paralysis seem to have taken over, events reverse. The open secret also recalls the simple, open facts of public life (a simplicity economists, for example, have had to strive ever harder to obscure). "Everybody knows," to paraphrase a recent Leonard Cohen song, the absurdities of public issues and political life; this is where all those call-in talk show comments about "common sense" come from, as much as from vestiges of Louis Hartz's civic liberalism. "You don't have to be a genius" to understand and solve this or that problem, Americans say—"this isn't rocket science." Surely, they are objecting to a politics they don't seem to be able to penetrate, but just as surely they have noticed how open the secret is; politics has become simpler, even as it has also become more dangerous and oddly harder to *do*.

The political mood I have just been sketching amounts to the background for the "dumb" character of politics Baudrillard has noticed; deep intelligence is not required to notice the workings of power, and motivation to *do* politics has little to do with information, its absence or complexity. The appropriate political response might not wish to "match wits" with a dominant discourse that wins all arguments (even the ones it seems to lose, but nonetheless are easily incorporated). As Baudrillard counsels us, the appropriate response to modern power may be to go inert, as much of the public evidently already has decided. A deconstructive strategy might not have to be as witty as our political science professors and expert commentators have made it seem. Perhaps the open secret serves as the sign that Americans have matured, outgrown the naive idealism that drove the 1950s. But this sign, too, turns metonymic; we can now understand that this was not a maturation at all, but a way to live with the bomb. The open secret is that the bomb is impossible to live with, and that the calculations about how to escape (for example, Willis's entrepreneurial attitude, discussed in an earlier chapter) are not helpful.

A politics of representation would be expected to have great difficulty with an era which turns "citizen" into object—calculator, entrepreneur, raw material for catastrophe—with such apparent ease as is now the case. (Conversely, the notion of citizen-as-political-author surely set up a situation in which the moves of power against the subject had an easier task.) Democratic theory has always assumed one direction of movement; the other direction—from power to attitudes—is equally plausible, as Edelman explained long before Foucault became important to American political thought: "It is therefore political actions that chiefly shape men's political wants and 'knowledge,' not the other way around. The common assumption that what democratic government does is somehow always a response to the moral codes, desires, and knowledge embedded inside people is as inverted as it is reassuring."[15] Every postmodern political impulse starts from some objection to this new direction of power, from a rejection of the reassurance Edelman cites. And after that change, all sorts of political action become possible.[16]

If Baudrillard has erred—failing to notice how close the actual workings of politics are to his discussion of the end of the political—we might ask how that error came about. And it might be appropriate to ask this most evocative and emotional of contemporary authors about the mood, the attitude under which his analysis proceeds. Baudrillard is, it turns out, all attitude—astral America, inert "majorities," cool memories. This is his version of an aesthetic political move. Politics with attitude.

In other words, this could be an issue of mood and attitude, of posture toward politics, rather than an issue of politics itself. Strategies are at issue here, to be sure; perhaps the issue is how we are to understand their "fatality."[17] The demise of depth analysis, representational mechanisms, truth-telling, and so on, could, after all, engender several possible attitudes: Baudrillard's melancholy, to be sure, but also irreverence, discontent, among others.[18] As Baudrillard understands, there is no theoretical basis from which to criticize his melancholic stance, but the same also goes for the alternatives. Each possibility stages reversals; Baudrillard has, after all, also deconstructed most of what caused the modernist artist to adopt a (seemingly) similar melancholy. As I have already suggested, the overall mood—encompassing either melancholy or giddiness—is irony.

We could focus on that attitude, rather than on the various political actions rendered inoperative by the attitude. In this age of communicated images, it is no longer necessarily the case that only the design of institutions and the choreography of official politics can influence citizens and constrain or create responses. Attitude, communicated instantly over our many networks, can now serve as a microcosm of response, a kind of governing mood. I have already been discussing the ironic as one such moment. Another possibility is suggested by Peter Sloterdijk's analysis of the cynical.

## Sloterdijk's Cynicism

Baudrillard's disposition toward reality — "the real" — is explicitly a cynical stance, one that disbelieves or deconstructs nearly every social commitment that has marked the age of modernity. While Baudrillard's work continues to explore that cynical mood, there is another place we might look for confirmation of that mood's status.[19]

Sloterdijk defines cynicism several times in the course of his long, often not-entirely-serious book. His "first definition" serves to introduce the mood: "Cynicism is enlightened false consciousness. It is that modernized, unhappy consciousness, on which enlightenment has labored both successfully and in vain. It has learned its lessons in enlightenment, but it has not, and probably was not able to, put them into practice."[20] Sloterdijk's central expression of cynicism connects it to the kynicism of the ancient Greeks, especially Diogenes. "Ancient kynicism, at least in its Greek origins, is in principle cheeky." This cheekiness is explicitly a response to impossible philosophical dicta, as I have discussed earlier in this chapter.

What Sloterdijk adds to Baudrillard's cynical stance — in addition to the aggressiveness and shamelessness of his project to counter melancholy with cynicism — is the (anti)history of a split in the cynical mood. Sloterdijk's definitions usually pull two ways; while cheekiness characterizes the best response to power in this cynical age, the exercise of power is also cynical, untouched by attempts to constrain it in traditional ways. Both power's operations and the culture toward which it is directed display cynicism; a cynical response finds new ways to refuse the manipulations of cynical power. (This double cynicism — aspiring to a very broad description of the age — is what gives life to the title's play on Kant's *Critique of Practical Reason*, playfully recalling his similarly broad ontology.)

The cynicism displayed by modern power is steadfastly skeptical about representative mechanisms, long since having abandoned the standards of "truth-telling" we would like to impose on institutions and other, less palpable structures. The cynicism of modern power gives Sloterdijk's book much of its rhetorical force; we recognize immediately that this cynicism—even more than the interests it serves—is what marks the operations of power in our age. There is a shock of recognition, reading Sloterdijk's evocation of the mood of modern power, which he calls "master cynicism."

> Modern cynicism . . . is the masters' antithesis to their own idealism as ideology and as masquerade. The cynical master lifts the mask, smiles at his weak adversary, and suppresses him. *C'est la vie. Noblesse oblige.* Order must prevail. . . . In its cynicisms hegemonic power airs its secrets a little, indulges in semi-self-enlightenment, and tells all. *Master cynicism is a cheekiness that has changed sides.*[21]

Sloterdijk understands the efficacy of master cynicism's response, in league with the potent, if false, remnants of the Enlightenment ("the idea that it would be reasonable to be happy" in "our gloomy modernity"). It is this possibility—that false consciousness in the service of power might prevail after all—that sets the stakes of the nuclear age, in Sloterdijk's view. "Atomically armed civilizations . . . are going through a crisis of their innermost vitality that is probably without historical parallel." In this period of "chronic crisis," it is demanded of all humans—not only the politicos who are accustomed to the ambiguity of important matters—that we "accept permanent uncertainty as the unchangeable background of its striving for happiness."[22]

Sloterdijk thus provides the best description to date of life with the bomb; life under security provided by this permanent uncertainty will be life against itself—life against liveliness. His use of Diogenes as the embodiment of cheekiness reminds us that this is not a new development, but the book's thorough discussion of twentieth-century European politics underscores the distinctiveness of this age. As I have discussed in earlier chapters, the bomb functions against longstanding human archetypes (such as the heroic warrior), and displaces any other metaphysics, becoming our "last, most energetic enlightener."[23] In short, Sloterdijk displaces all prevailing, serious, and straightforward totalities with fake ones— kynical ones, informed by enlightened false consciousness, at war with the cynical totalities that oppress and destroy.

The great strength of Sloterdijk's argument is how easily it finds a place for nearly every contemporary political event within its framework, its universe of cynicism and cheekiness. The tiresome topic of "politician's lies" is rejuvenated as master cynicism: a way of life, rather than an exception or transgression of an otherwise acceptable and functioning, enlightened order. Likewise, the solemnity of the critic—whether the liberal or the critical theorist—is put to a test when we suggest that the very problem may be the impossibility of what they have been insisting power do. "Communicative competence" disappears as a signal concern when we understand that power is already stunningly competent at cynical manipulation. Baudrillard's longstanding quarrel with the validity of "the realm of the real" finds an important new ally; the spokesmen for the real have long since adopted a cynical mode, and their opponents can trace their cheeky replies back to the same ancient source the rationalists claim.

Perhaps most important, Sloterdijk's analysis helps us clarify the postmodern political impulse. There is a tension in Baudrillard's work, between the declaration of the end of politics, on one hand, and the close parallels between his work and the political impulse I outlined earlier in this chapter. As it turns out, Baudrillard's analysis is not deployed against politics per se, but against all claims to reality and truth. This is the context for Sylvere Lotringer's remarkable response: "Jean, your theory is too true to be good!"[24] The demise of Enlightenment-era politics is such a dramatic event that "good" and "truth" are radically inverted, and we might as well speak of these inversions as an end of politics. Within Sloterdijk's project, Baudrillard's philosophical assault on politics is confirmed, and attains its (paradoxically) appropriate, political position. The best response to the nuclear age and the triumphant era of master cynicism would be an ongoing *political* enterprise — cheeky, savvy about the crisis of representation, and committed to political projects just at the moment they apparently become futile.

Sloterdijk asserts the efficacy of kynical politics, and that part of his argument is not as persuasive as his diagnosis. The cheeky response takes apart liberal modes of opposition; it may or may not also take apart power. That is for political action to decide, and there are good reasons to suspect — as Sloterdijk admits — that it may already be far too late. But this possibly excessive confidence in the efficacy of the kynical response would be a minor error, when juxtaposed to Sloterdijk's stunning diagnosis of our age. The cynical mood provides confirmation of the postmodern condition, in a way more fully matching the dimensions of the condition, in a way that Lyotard's more direct (if less political) statements cannot match. And there are reasons, as well, to think that the events of 1989 and 1990 have now — just at the point of the disappearance of politics — finally captured the ground for what now becomes politics itself, creating the grounds for the response Sloterdijk could evoke only through historical analysis and hysterical example.

## 1989

Nobody doubts that extraordinary events happened in 1989. On nearly every continent, situations that had seemed permanently stuck in a miasma of totalitarian control simply began to unravel. While this could not be mistaken for a simple outbreak of "the good," nobody seemed entirely sure what had happened. Most intellectual analyses emphasized the details; there were new actors and new institutions to know.

Although it is too early, at this writing in mid-1990, to declare very much, I would still venture a tentative position. We have entered some early stage of the culmination of the nuclear age I have been describing, and its refolding into an age best understood within the position detailed by postmodernists, especially Derrida, Baudrillard, Sloterdijk, and Foucault. The age is marked by a new and unexpected prominence of certain themes and processes. "New rules" emerge (through the mechanisms of a master cynicism that has consolidated its position for decades with only sporadic difficulty). This new, late modern power now, belatedly, meets its opposition, a counter movement of extraordinary, if intermittent and unreliable, efficacy. The features of this encounter between late modern master cynicism and postmodern kynicism should not surprise us; the reversal of foreground and background is not a simple procedure. To understand this era, then, we begin as every generation of intellectuals has begun when faced with inexplicable change; we look to emerging and unexpected developments in the rules that govern the world of events. We need to investigate precisely what rules have changed, how the new rules work, and what the implications of these changes will be.

We can provisionally identify some of these new rules, remembering that several of them are the simple reversal of predominant conceptions of space, time, and event. I would summarize these as: (1) the new role of pace in political communication; (2) the relative deprivileging of territory as defined by geopolitics; (3) the displacement of ideologies by communications and newly potent commodity relations; (4) the displacement of sight by spying; and (5) the kynical and fractal realignment of such oppositional forms as the mass demonstration in response to the crisis of representation and new domains of power. What follows, then, is a microcosm of the emerging world and the opposition that has both promoted that emergence and is now necessary, in this provisional analysis, for further opposition.

## Pace and Political Communication

The instant communication of political reports and demands becomes a phenomenon in itself; what had previously been in the background, a technological curiosity, now is well enough developed and integrated to become a phenomenon in its own right. Feeding on itself, this pace soon becomes momentum, as any football announcer knows. Force is transformed from the realm of weight and physical power to the realms of expectation, change, and the adrenaline rush of political action chronicled in American electoral politics by Hunter Thompson.

The nuke was the first important sign of this emergence; the telephone, automobile, and aircraft were subsumed under the (triplet) signs
of the nuke, the tube, and the microprocessor. Our Fordist culture has
long read these emergences as harbingers of yet new products to come:
neutron bombs! flying cars! a mutant television-telephone! hologram
video games! celullar FAX machine-mojo wires! bioengineered life forms!
No doubt, some of these products (and many others) will emerge; there
are, after all, many Fordists among us, and they have huge reserves of
energy and power. But the actual "end" of their enterprise is now visible;
no matter what the vehicle, the effect is defined by the pace of advancing
developments. Successful actors (oppositional or not) will be those who
can function at speed. That is the emerging rule; expect a gallop of
events, and take seriously that enduring Deaverism—the edict always to
"stay ahead of the curve."

### Geopolitics Reoriented

Territory suffers as the central goal of politics, even while space emerges
as a newly important figure of speech, displacing light and observation.
Germany and the Middle East combine to teach lessons; the combination
of Germany—never imagined before it happened—became possible because territory at least recedes from its customary role as the most important border of events, the container of last resort and judgment. The
pace of travel (of weapons, of goods, of information) renders several sets
of seemingly solid borders archaic. This possibility—unimaginable until
the reunification of Germany became a fait accompli—soon becomes almost obvious. This is not a universal development—all sorts of borders
retain their ability to define and orient political disputes. As the Middle
East reminds us, resources are still distributed territorially. But the oil that
now serves as the most coveted resource fuels movement, and that motion
has already assumed new forms, as information takes over. The remarkable spectacle of American broadcasters in Baghdad, even as Iraq pre-
pared for war and then fought against the United States, marked a border
dispute more important than the Iraq-Kuwait tiff itself. Modern communications "covered" the territorial dispute; eventually, we now suspect,
the coverage will win.

Knowing that territory and oil will both pass quickly into the realm of
the forgotten, we can still go to war over them, knowing full well that this
will not "end all wars." This is cynical military action, incapable of re-
suscitating the warrior; throughout this hugely popular military exercise,
enlistments actually went down. The role played by the United States in
Panama (during late 1989) and the Persian Gulf, just a year later, may yet
turn out to have been a brittle, frenzied attempt to preserve a role for ter-

ritory, after the demise of the Berlin Wall. Or, if the United States insists on playing its hyper-militaristic role anyway, at least the difference between our antagonistic role and Europe's newfound unity will be more difficult to miss.

## Ideologies Displaced

A distinctive feature of these transformations is that they elude capture by the existing ideological apparatus. No matter how hard the capitalist West tries to proclaim a victory of its own ideas and institutional arrangements, it becomes clear that the actual victory must be awarded to change itself, to media image, and to the ironies and cynicisms that gather in this new arena. Many of the essays in this book have argued that a basic incoherence—not precisely lies or manipulations or confusions—necessarily informs our political culture. This is not the grounds for ideological victory (or the vindication of critical theory either, for that matter).

As Timothy Luke has recently argued in *Screens of Power: Ideology, Domination, and Resistance in Informational Society*, it is the communication that now alters ideological questions, displacing any critical attempt to impose the reverse. Any Western "tradition" that could be said to have "triumphed" had long since already been transformed into an ironic, floating "teletradition." In one very useful example, the "return to traditional values" in American society is driven by an unconcealed commodification of image, a development that simulates a connection between family values and "telegenically sold products," in a format Baudrillard captured long ago.

> In consumer society, Jesus or God, like Wonder Bread or Geritol, is sold telegenically as a product that heals bodies, mends filth, and changes attitudes. . . . Taking this logic to its ultimate conclusion, Jim and Tammy Bakker even built—before their fall—an amusement theme park called Heritage, U.S.A. . . . which was billed as a "Christian Disneyland," to round out their televangelical consumer product line and prove that fundamentalist faith can be "fun."[25]

The trap, of course, is that a critique of this commodity culture itself becomes commodified. Luke shows how a newly invigorated neoconservative movement has been able opportunistically to marginalize critical intellectuals who misunderstood the new conditions under which they worked. Luke's recommendation for a reconstructive opposition, using "alternative media" outside corporate and bureaucratic control, shows how Foucault's specific intellectual looks to the American scene.

> Critical intellectuals could help revitalize many forms of resistance within the everyday vernacular dimensions of politics, language, society,

and culture at the local, municipal, or regional level by creating new counterinstitutions. . . . The intellectuals' contributions of critically grounded "good sense" might recombine with the popularly oriented "common sense" of other individuals in their local communities to resist the present administrative regime from within. Fortunately, a permanent resistance to these ideologies and their domination already exists among many diverse audiences, who are puzzled by—and question—many of the images they scan on the screens of power.[26]

Acknowledging that such responses "could all fail," Luke has still established the new, marginal ideological position from which we could yet move. The telecommodity becomes the new arena, not vindicating any ideology so much as forcing a reexamination of intellectual roles that have always underpinned ideologies.

## Ways of Seeing/Ways of Spying

Sight and illumination no longer function as they did throughout the modern era, but they do not go away entirely. The spy becomes an important microcosm of the era, emblematic of what power does, emphasizing surveillance and the play of power's seeming absence, on the one hand, and citizen self-control, on the other.

The spy's currency is a particular kind of information, which exists only on the condition that it be acquired cynically. The spy thus exists at the point of intersection between Sloterdijk's cynical age and Foucault's disciplinary one, showing how these two are linked in the postmodern terrain. The reversibility of the spy is her most basic protection; there is always the chance that the spy has already changed sides, and awareness of that fact conditions every encounter with the spy or the information she produces. But despite that reversibility, nobody doubts the spy's (at least sporadic) efficacy; in that sense, the spy establishes both the condition for postmodern politics as well as the authority that politics will have to oppose.

Matching the spy is a tricky enterprise; spy vs. spy is, after all, fully within the spy's game. One more reversal is always possible, at least until the spy opts out of the game and takes public refuge in one capital or another, quickly telling secrets that just as quickly begin to devaluate as the other side adjusts. But the linked instabilities at the center of the spy's project are not necessarily debilitating. This is the genius of surveillance, as Foucault established when he explained how the spy can even make use of absence, by distributing signs of possible presence (as in the Panopticon). It is, as well, a feature of espionage deployed with phenomenal success by Richard Nixon and the various "red hunters" of the 1950s; while the left earnestly complained that various accused spies were innocent, the conservative "master cynics" took full advantage of their early discovery of the political utility of surveillance.

The left (always enamored of the real too long) finally developed a response, in the almost complete openness of most civil rights and antiwar groups of the 1960s. In radical meetings of that era, activists frequently addressed the spies they assumed were present, somewhere, in their audiences (or on their phone lines, or in their mailbox). But this response was only marginally effective; by then, the spies had found new ways to operate, using their ability to create identities in yet another way. J. Edgar Hoover and the generation who followed him created an identity *for those they watched*. Spying became an element of the production of political enemies. That this iteration of strategy and counterstrategy can continue, as it has, is perhaps the best confirming evidence of a politics-without-essence, establishing the possibility of postmodern responses. In a sense, George Bush announces the era of surveillance as effectively as Reagan had announced the hegemony of the image. With Bush, we have the master spy, out in the open now, playing electoral politics. And still the spy's game perseveres, still somehow almost impossible to expose. The spy rules.

## Oppositional Forms

Even if, taken singly, the works of each of the postmodern analysts I have been discussing fail to produce space for opposition, the accumulation of their work now provides just what they resist. Baudrillard seems to have chosen a point just beyond any imaginable politics as his point of departure, and that choice conditions whatever political commentary appears in his works. (His ostentatious grumpiness over feminism, emphatic in *Seduction*, is only the most recent example.)[27] Sloterdijk politicizes knowledge in a way directly translated to political action, but is, so far at least, working well within the historical period of late modernity. Foucault was working on these themes late in his life, but they were left undeveloped.[28]

It seems likely that the next major project of postmodern analysis is to reread the political forms of opposition that have succeeded in recent years. The apparent opening of political space is hardly a rebuttal to Foucault, Baudrillard, and Sloterdijk; indeed, to formulate the question in that fashion is to show how absurd such a conclusion would be. Opposition to new forms of authority, propitious use of the speed and fractal character of change, the sometimes frivolous attitude toward the ends of radical action—all of these were evident in 1989 and 1990, and each confirms the possibility of postmodern oppositional tactics.

At some point, the combined effects of speed, surveillance, and deterritorialization produce an entirely different kind of political change. Baudrillard has described this new transformation as fractal, and the name fits. Recalling the new fractal mathematics, this mode of change is depth-less but also extensive, incremental but also instantaneous. Transformations now array their effects throughout the "body politic," all at once.

> After the natural stage, the mercantile stage and the structural stage, comes the fractal stage of value. To the first corresponded a natural referent, and value evolved in reference to a natural use of the world. To the second corresponded a general equivalent and value evolved in reference to a logic of merchandise. To the third corresponds a code and value unfurls itself in reference to an ensemble of models. To the fourth stage, which I will call the fractal stage, also the viral stage . . . there is no longer a referent at all. The value radiates in all directions, filling in all interstices, without bearing reference to anything whatsoever except by mere contiguity.
>
> At this fractal stage (which one can relate to the fractals) there is no longer any equivalence, natural or general. Also, there is not any law of value as such, dialectical or structural. There only remains a sort of epidemic of value, a general metastasis of value; a sort of proliferation and problematic dispersal.[29]

The recent return of old ethnic confrontations in the Eastern bloc teach a lesson unnoticed so far; it is the United States that subsumed—fractally transformed—its own ethnic differences (with the obvious and important exception of differences seen as "racial," namely *vis-à-vis* Native Americans and African-Americans). Totalitarian control (based on totality, overt and modernist force) failed; the American model succeeded. As Baudrillard suggests, there is no obvious judgment to be applied to this process (or within it); each explanation or evaluation is reversible, and is, in any case, quickly outdated.

The transformations of 1989 and 1990 fit Baudrillard's new fractal category. Each event was as instantaneous as the famous FAXes transmitted to machines controlled by protesting Chinese students. The political upheavals were both complete and incremental; ideology was given much less attention than spectacle, but governments fell. The exemplar, no doubt, was Czechoslovakia, where even a few days before the demonstrations began nobody anticipated change. In the blink of an eye, a playwright was head of state and an expatriate artist was the chief diplomat in a Washington embassy known for spies and listening devices. Even the "counterexamples" of 1989 and 1990 confirmed the new postmodern condition; George Bush's Panamanian adventure seemed quaint, anchored only to the Drug War spectacle in the United States. Everybody knew that the Tiananmen catastrophe only delayed for a few years the transformation of a remarkably durable Chinese government. And, at least in its early days, the deployment of U.S. forces to Iraq in 1990 drew attention to the startling reversal of a very predictable Soviet diplomatic stance in the area, more than to any resurrection of the warrior (who was, in any case, wrapped in a plastic gas-suit, absurdly dressed in a desert, practicing for the onset of biological—viral—war).

In Berlin and Beijing, especially, the televised display of the kynical attitude became the central political act. Nobody much questioned, anymore, the new preeminence of the street demonstration. Advice to "slow down" had been a mainstay of the counterrevolution in the United States during its own outbreak of demonstrative politics a quarter century earlier, but was seldom heard in these instances—the pace of the events was the point. Ideology receded; "the people" were a street event, not a construct of representative interest and will. No longer a representation of "the people," the demonstration had not been robbed of its efficacy, and in fact worked better than ever, pushing the pace of events to giddier levels yet. The demonstration is only the arena, the microcosm that accompanies the transformation. It is the metaphorical street that was captured, not the physical one (where one would, say, prevent the delivery of parts to an auto plant on strike). Here, the street was at issue because the show wouldn't fit in a proper television studio.

On the other hand, couldn't this be simply an excessively dreamy fable of deconstruction? No doubt, but in the wake of a year of extraordinary political activity—in the wake of a nuclearism everyone accepted as the end of such politics—events proceeded that demand fables, as well as detailed coverage of specific elements. Gorbachev and Reagan had finally issued forth an unmistakably postmodern era, a triumph of deconstructive strategies. What else could the removal of the Berlin Wall mean?

Less than two years after Tiananmen, of course, it appears likely that each of these transformations had inexplicably paused, or even reversed direction again. The U.S.S.R.'s unmanageability overwhelms whatever might happen in Eastern Europe. The U.S. military role has been reestablished, while contention pervades South Africa and the Chinese regime is evidently in firm control. A fractal counterrevolution, it would seem, could be just as complete and sudden as its revolutionary referent.

For the theorist, if not for the subjects of oppressive regimes, this outcome may be preferable. We have no generalized "new day" here; we have, instead, a broad change in how political change happens.

# 7
# On Lastness: Nuclearism and Modernity

Of the many paradoxical dimensions I have been associating with the nuclear age, perhaps the most important one is that day to day life continues, frequently forgetting the nuclearist transformations that really are impossible to ignore. Politics finally went orbital; encircling us in the most general way possible, but at a sufficient distance that it seldom intervenes very directly in our picnic and softball game. Most reminders of nuclearism thus take on an ironic or absurdist quality: "How can you call that pitch a strike when nuclear war could break out any time now!" Certainty vanishes in this context; there is no way to extrapolate a plan for action from our available ideologies.

Concluding a book on such a situation is no small consideration. Faced with such a dilemma, I looked for a Situationist way out; what was at hand, available to be appropriated against itself? An invitation to deliver a public lecture at the university where I teach was at hand, so I picked it up; this conclusion is offered as an outrage, a little absurdity.

The lecture series, a long standing tradition, is called the "Last Lecture Series." The premise of the series is that the lecture should simulate the last lecture one would expect to give, at the current stage of one's life. Given my oft-voiced commitments to be especially attentive to political power and the subtle way it works in the structures closest to where one is, I decided to risk some melodrama by talking about the Last Lecture Series itself, and to use that as my little book's conclusion.

The Last Lecture Series is not a frivolous institution; the decision of a junior faculty member to treat it frivolously was not obviously defensible.

This is one of the ways our small, public institution uses to imply the life of a small, liberal arts campus. But more than gentility has driven these lectures. That title—"the Last Lecture Series"—implies a capping up, a chance to reflect, perhaps a grand instant replay. What is more, this distanced reflection—given credit precisely because of its distance and detachment—is an enduring, fundamental characteristic of social analysis in our culture. This is, after all (a phrase favored in such lectures), the sort of distance Plato quoted Cephalus as praising in the Republic.

Ever since, Western intellectuals have been very serious about this idea of a privileged narrative, delivered in view of a whole life. But even though the idea of the heroic intellectual began to be impossible in the last century, the models of professionalism, career, and project helped the distanced narrative maintain its privileged status. John Dewey, for example, kept this possibility alive in his arguments for a reasoned, scientific pragmatism. In his schema, one assembled a life and, thus, it would be possible to recap that life.

Deconstruction aimed squarely at that narrative, undermining the possibility of narrative itself, and also making perspective a target. Increasingly attentive to language and disjunction, some social scientists have begun to understand that even the words with which we describe our lives are never as clear as we hope they are. Whenever we indulge in hopes for a direct line to the real, language keeps showing up instead. And deconstruction keeps going—defending that lack of clarity and extending it to authority, thus finding some possibility for political action. At least, that is the case I have been making in this book.

In part, that position was built on ground taken earlier; it was Merleau-Ponty who dramatically identified the role of contingency in what otherwise might be mistaken for historical narratives. He privileged the role of ambiguity as a crucial attribute of history, and his position undermined all mechanistic approaches to politics, whether liberal or Marxist. After Merleau-Ponty (who wrote at the very beginning of the nuclear age), political action required an aptitude for dealing with pervasive ambiguity; conviction receded from its role as an essential political characteristic. Finally dispensing with the traditions of Cartesians (a heritage for both determinist Marxists and liberal "originalists" like Robert Bork), some of Heidegger's readers understood that the ambiguity itself might be the important issue, more important even than that which was ambiguous.

The typical political translation thus became the movement from ambiguity to positive claims. Some leap—identified by Camus as the jump from the fact of existential individualism to the presence of some collective—emerged as the characteristic political subtext.[1] Modernists in art and politics may have relied on the heroic leap which one could infer from the existentialists, but a more radical inference would emerge. One of the accomplishments of the postmodernists is to explain that this translation from ambiguity to action has not only been possible, but has been characteristic of political radicalism for some time. That is the importance of Foucault's praise of political localism; already, even without acknowledging it, effective politics had already been partial, mobilized to challenge truth and emphasize multiplicities.

In a sense, then, something called the "Last Lecture Series" necessarily would occur at the fissure between two very different realms. This word "last" formed a boundary in the border skirmish between poststructuralism and the liberal humanism of the modernist age. Like all borders, this one forced a test. How could someone convinced of the unlikeliness of certainty and perspective accept such a final, detached, and still powerful presumption as the integrity of a last narrative? This could be one of those (fairly unusual) situations in which the context or setting might be so determinant—even in an intellectual environment—that it exposed its strong (but usually hidden) role. Or does "last"—that final, unambiguous perspective—prove that ambiguity can be fought on the grounds of tradition and values? That is, does the fact that we can write, publicize, and deliver a "Last Lecture" successfully undermine the nihilistic role played by ambiguity?

To do the lecture—or to live in the nuclear age—I had to contemplate *last*. It was not much of a chore to delegitimate this very common word, beginning the process of robbing its finality, its certainty. The dictionary definitions of last include "remaining after all the others." The runner in a cross country meet who finishes behind everyone else is said to finish last. He or she remained after all the others. But the winners in a tag team wrestling match on cable TV are the ones who remain, the ones who last. The issue, of course, was what such little tricks accomplish. The devices of deconstruction seek to delegitimate the subtle, contemporary forms of authority that gather efficacy at this late stage of the modernist subject, showing that their ground is shaky and insecure. The ambiguities of language—its subtlety and slipperiness—are not simply a regrettable episode in an otherwise lucid world. Deconstruction exploits those slippages of meaning, makes puns and puzzles, to demonstrate how thorough and inevitable this ambiguity is. So, the "last" on a shoe is the sole; following singer Paul Simon, I could talk about diamonds on the last of her shoes, perhaps an Imelda Marcos joke.

I remember being fascinated, as a child watching the Gillette Friday Night Fights on early TV, by the word EVERLAST, which all boxers then wore on their trunks, like a belt buckle. I learned it advertised athletic equipment, the boxing gloves and shoes the boxer wore. More important, but much later, I would learn why it was a fascinating word. In part, it was an advertisement that someone *wore* (an oddity then, commonplace now). The word EVERLAST represented a commodity, and the boxer's uniform added that word to a body; it was an unusual display then, and shocking.

EVERLAST was also a contradiction. Big, bold, block letters—very strong—promised to last forever. This word seemed to have a very literal power. At an early age, I had already learned that words were capable power, and this seemed to be a special case. Here was a strong word on the body of a young man who seemed strong enough, but was actually very vulnerable. In the ring, one of these righteous, EVERLAST-ing dudes would drop tonight. In life, very likely both would drop. Undereducated, ill fit for changing societies (which had blithely promised that power and notoriety were coins of the realm), they were tragic heroes, as Martin Scorsese's great film *Raging Bull* attests.

If putting EVERLAST on a boxer is odd, putting the word "last" on an institution is even stranger. A Last National Bank would be a definite curiosity. A street named Last Place? And here, a Last Lecture Series. The institution of Lastness. After such a lecture (if it were "real" and not the reenactment on which this lecture series is based), even the speaker would be quiet. But having survived, having "remained after all the others," the speaker might have insights to impart. A last lecture indicates finality, and by doing so, it also implies—and confirms the significance of—a career that is eventually ended with the ceremonies of transition. Standing on the cusp between silence and greatest insight, the elder of the clan would be in a place very privileged by the customs of career and profession, perhaps even linking those customs to older traditions of deference to elders, seniority, and perhaps the idea of tradition itself.

Given the possibility of nuclear annihilation, we have added a new genre of Last Statements. In a scenario reenacted continually in popular art forms, the blast produces (miraculously) *articulate* recognition, honesty, and regret. The blast and disorder are placed into the service of clarity, in the hope that horror over such wasted Last Thoughts could be reenacted before they happened, in fictions, so as to prevent that awful event. But precisely the possibility of such a communal demise means that the idea of Lastness has irrevocably changed. The presumptions and structures that make a Last Lecture Series possible are now vulnerable; we have to grapple with them. But I obviously don't believe that this requires the dismal, desperate attitude adopted by some antinuclearists. Rather than ending our prospects, the demise of Lastness opens other possibilities.

The theorists discussed in this book present continual reminders that it is still possible to move in some positive way, even after a social death—something quite beyond the death of the social—is conceivable, perhaps even likely. We might begin to establish that positive possibility by reenacting an obsession with Lastness, unashamed and engaged. Surely, other speakers in the series must have addressed Lastness, since it is such a provocative topic for us mortals. But I decided to talk of nothing else, to deconstruct the conceit and to do so with an attitude that would test the morbidity of most antinuclearist warnings.

Lastness ... Finality ... Totality ... These things fit together. To speak at the end is (has always been) to be able to speak the truth. That is what Plato's Cephalus was suggesting, though Plato quickly dispensed with the old man. We grant special credence to someone's last words, even if we also quickly dispense with such a speaker. Deathbed confessions are considered so true that they'll stop any cop show plot flat in its tracks.

We expect to arrive at some point where we will have a last word. Time extends in increments, and at the end of the increment someone sums up the events that transpired. We might call this the Walter Cronkite theory of Lastness, after its great exemplar. Everyone's Uncle Walter always ended his increment (the evening news) with the statement, "and that's the way it is, this 22nd of November, 1963." The day had happened, and Walter had recapped it.

At a certain point, a few years into the Vietnam War, Walter's closing cue became, improbably enough, a laugh line. Right, Walter, just as it is. The exclamation "What it is" (sometimes a question, sometimes a statement) became standard black English. A young black poet and performer, Gil Scott Heron, began a new talking blues tradition with a little ditty called, "The Revolution Will Not Be Televised," and later satirized the excessive solidity of Cronkite's message, in a reference to "Walter Concrete and the News."

Different people had different reasons for laughing at Walter, and some of those reasons were untenable. A significant number of my peers thought they knew *what it is*. They were sure Walter was covering up what could be revealed through some ideological lens, even if it remained unseen by the CBS eye. That critique failed; it was simply too obvious that the ability of anyone, even Walter, to recap with that kind of finality no longer existed. The lenses we had used to see (and show) it like it is had shattered.

As Albert Einstein understood, the bomb really did change everything, even if it would take a while for that change to emerge fully. Walter's line was funny, all of a sudden (even though we'd seen it hundreds of times before without thinking a thing of it), because nobody could sum up the day's events with his very convincing calmness and rationality once the bomb was among us. No, after Hiroshima (the name of a town, now metonymically extrapolated to stand for the human condition), all statements about how the world was would have to be delivered in a way that at least acknowledged the possibility of a loss of control, or of control producing that final fall.

Perhaps if Walter sobbed one night, took off his shirt and danced on the table the next, mumbled incoherently the next; perhaps then we could agree that that was, indeed, the way it was. Such a possibility yielded a successful movie, *Network*. In "real life," during the campaign to determine Ronald Reagan's successor, Dan Rather did great things for himself, the Democrats, Bob Dole, *and* George Bush by merely hinting at the possibility that he might break down and get truly pissed at the vice president.

In short, the presumptions that most of us have lives to live out, that our lives would be followed by other people's lives, that our rationality and humanity would gradually make things clearer, or even improve them—all were undermined forever by the bomb. A narrative was disrupted, replaced by a fracture. Have your expectations, hopes, and commitments, the bomb said, but always remember that one wildman, one moment of weirdness, could cancel them out.

It is true that there have been attempts to conjure the end of the world before now. The Oxford English Dictionary recounts that even this word "last" has long had such a connotation. But our finality is different. We know that we've decided to organize our lives under science, and we know that science and technology can touch the entire world, ending it.

The bomb takes us completely into the realm of fiction, myth, and unreality. Derrida has made this point quite convincingly. We are constantly told that nuclear war is unthinkable, that it is impossible to contemplate. But, Derrida argues, nukes are far from "unspeakable." Derrida's breakthrough is his explanation—utterly disconnected from apologies for nuclear managers—that nukes aren't unspeakable, but are, on the contrary, uniquely spoken, or textual. By "textual," he means that the politics of nukes are "literary creations" in a sense that other political questions have never been. Information, communication, codes, and decoding are at the heart of nuclear politics. When we talk, language comes out, and very often those words really do "have a life of their own," as we sometimes say. Try as we might to say "what we really mean," we are speaking a language that has huge inertia. People were speaking it long before we were born. You could even say it "speaks us"; it creates our context, it is the range of possibilities within which we move.

All of that is becoming commonplace, at least among intellectuals. More serious is the question of democratic politics. A political culture in which the participants (at least some of them) know the culture exists only in words, anticipation, and image will be quite different from other varieties, in which specific sets of facts and objects have been privileged. Nuclear war not only is not a fact, it can never be a fact. What is more, this emerging politics is ahistorical—if you are convinced that history provides the basis for the solution of our problems, you will have to deal with the nuke, which escapes history. As Derrida notes, "A nuclear war has not taken place: one can only talk and write about it." Or "Nuclear war has no precedent. It has never occurred, itself; it is a non-event. . . . Some might call it a fable, then, a pure invention: in the sense in which it is said that a myth, an image, a fiction, a utopia, a rhetorical figure, a fantasy . . . are inventions."[2]

This fable-like quality—this fabulousness—helps us to understand some aspects of antinuclear politics that might otherwise seem extraordinarily out of place. Many commentators have noted that nukes, of themselves, do not stand out as a risk. Coal damages more lives than do nuclear reactors, even after Chernobyl. Many soldiers and civilians are killed by machine guns and primitive explosives. Chemical and biological warfare is, in some ways, more fearsome than nuclear war. That is to say, there is a disproportion of fears to risks evident in this discourse. Some apologists for the nuke find this disproportion evidence of the looniness of their opponents, but that is too convenient an explanation to be taken seriously. On the other hand, some antinuclear writers point to these extraordinary fears as demonstration of a Jungian Fear, a pattern that traces throughout civilization and history, providing a unity, albeit a negative one.[3]

This odd situation bespeaks some kind of inarticulate acceptance of the fabulousness of our times. One needs a fiction, a symbolic body, to deal with modern life, in which so much rides on myth and image. If the symbol doesn't exactly fit the "reality," so much the better; that only "proves" there is some stronger power, lurking behind the risk-assessment calculations. When an unquestionably successful president uses Star Wars to quell opponents, then turns out to have been seriously infatuated with astrology, the benefits of hard, technocratic, "true" analyses begin to crumble. The choice of a fictional risk, pressed disproportionately to its "real" references, is a strategic choice, an intervention that polarizes the advocates of reality, exposing their hubris. It is playful.

Of course, there are other ways of discussing the demise of the real at the hand of powerful fictions, all of which I have been calling the "end of lastness." I could note the demise of Christianity as a community-wide myth that had power and truth. But the plot is so similar that my additions would simply be redundant. Lastness went south. The idea that at the end of our increment, at the end of our career and life, we would be able to sum up the events and successes and failures, now had a big, glowing hole in it. That's serious. We had learned to comfort ourselves, knowing we would have that last word, at some undefined future point. Someday, we'll look back on this and laugh. That reassurance steadied the keel.

Even knowing when that last lecture will be is a problem. No contemporary speaker can presume the Socratic luxury of distance by virtue of old age in the way most speakers have throughout history. The somber, responsible mood of last is a sign that misfires, referring to no privileged perspective or knowledge. Without that sign, Lastness would hardly form a coherent phenomenon. If we took it seriously, there would have to be some joking aside; the mood couldn't hold.

There are other ways to analyze this unexpected absence: the disappearance of the last. Social scientists claim to have learned a few things about cataclysmic political change, and one of the things they know very well is that real political change happens when expectations are rising, not when hopes are low. That rule of thumb has been demonstrated, over and over. Those who are actually starving rebel less frequently than those who have seen how to end their hunger, or those who expected hunger to end when it did not. This analytical rule corresponds very well with our commonsense politics; a critique must be accompanied by a proposal. Naysaying changes nothing, as the saying goes.

But neither political science nor common sense is notably effective in the face of the nuke. A massive exercise in self-delusion will be necessary to raise expectations, putting not only the problem but also the solution into the realm of the fictional. This is not a partisan insight; if you happen to take deterrence's peace-keeping abilities seriously then its success doesn't raise realistic hopes, either. It simply means that surveillance will be with us forever. Deterrence and nuclearism are now the classic excuse for surveillance. To repeat: whatever the particular political position, party, or ideology, delusion (fiction, metaphor, image) will henceforth be more central to it than facts, objects, and certainties. Reagan's famed vagueness with the "details" and even the broader circumstances of administration is not the "crime against the nature of society" some of his critics seem to think it is. But the qualities of his performance can still be exposed and criticized.

Political action can still proceed, even in lieu of higher expectations, objective analysis, or appeal to common consensus based on clear community values. By not recognizing this possibility, liberal antinuclearists have remained wary of some of the styles and attitudes that have, in fact, served them best. The heavily responsible, even dismal, attitude seems necessary; distracted citizens do not "want to know." Indeed, they are presumed to want to forget. Warnings and horror stories, then, are the liberal reminder that the escape to hedonism and consumption is a vain dodge. The privileged liberal form becomes the stern lecture, the sermon.

The appeal to terror—in Helen Caldicott's work, in films like television's "The Day After," and even on bumper stickers and buttons—has been a main staple of antinuclearist politics, a sober, anti-ironic terrorism of images. This is only the most recent, most desperate version of a political stance familiar in this century. Liberals, conservatives, Marxists, Christians—all felt, all along, that they were dealing with the *real*, with reality. The gravity of that encounter meant that public life would be formed in competitive narrations of sober, somber articulations of the necessary, the universal. As the real becomes less and less compelling a mediator of disputes, the mood turns desperate. If the horror stories fail to hold attention, the logic of the situation dictates a negatively nihilistic diagnosis of "human nature." Citizens who seem oblivious to the strongest warning call must be in the mold of Nero. The requirements imposed by the real—not the deconstruction of those requirements—are the source of nihilism.

The postmodern analysis is more compelling. Rather than a nihilistic subject, somehow already burned by the nuke, we have subjects constituted in new ways. Political authority is still explicable, even if shared convictions about reality become untenable. In a monarchy, subjects just *knew* that the king or queen embodied political authority—power was vested in the monarch. Then, the antiroyalists we now call classical liberals just *knew* that a scientific analysis of nature would liberate them from those old fashioned monarchists. Then, the radicals who followed Marx just *knew* that the science of historical materialism represented reality, promising a future revolution that would dissolve political authority.

These last two form the base of modernity. These two great opponents—liberals and Marxists—have shared a privileged certainty, which both obtained from a shared, scientific base. The term "postmodern" implies that the era is ending, but it hardly signals the demise of political power. No longer externalized in the body of the king or relations of production, or nature, power becomes internalized. We started to create a new self, a new set of habits and practices that we use to identify ourselves as individuals. We become calculators, seeing problems as opportunities to weigh costs and benefits, knowing very well that any other kind of motivation will be suspect. Still, we continually find ourselves in situations that demand that great decisions be made before we quite know the categories. Calculations require facts, but politics continually presents situations without facts. That should tell us that we are not calculating machines, or at least we aren't very good ones. And such contradictions begin to promise a politics.

We could, at least, protest against the kind of selves produced in the modern era. That politics would be cultural, its standards more aesthetic than scientific. It would be an odd politics, because it would not promise answers or any such certainty. Solutions, whether mathematical or chemical, would become less important as political metaphors, and, contrary to the words of deconstruction's many critics, this would not be the end of politics. Examples of a better politics abound, available to anyone who has abandoned the "solution" metaphor. Indeed, there exists an argument that makes it necessary, not optional, to identify "solution" talk as metaphorical, giving the fundamental option of discarding that metaphor to anyone who took all this very seriously. That is to say, identifying "solution" as a metaphor (and identifying other metaphors) was not an evasion of the challenge — what's your solution? — but the main work. That which must be done to understand where we're at.

There is an important implication from that decision. The political approach I have in mind would understand that power continues. Any political stance based on rising expectations sometimes exaggerates promises. Revolution, transcendence, liberation — all those are dubious promises, producing a falsely certain future in whose name action is justified. Postmodernism, then, clears ground for political action — not acquiescence — when it demonstrates that power will not somehow be rendered null, but will be the context of ongoing performance, exposure, release, and recapture. The demise of promises of generalized liberation is hardly the end of politics. Nor does that demise even signal that politics has become unappealing, a matter to be avoided. And it certainly doesn't mean that liberation is now located in some other area. There *are* still struggles to expose oppression and discipline and surveillance. As power is increasingly deployed in ways that do not oppress or dominate, but create us as power's subjects, inscribing its practices into our disciplines and habits, strategies of response will change, not disappear. And the new responses will share an identifiable pattern. Successful critique henceforth will come from those who can make what had seemed real, before they addressed it, now seem thoroughly problematic. We enter the era of the puzzle.

Faced with the critique that he had somehow ended politics, Baudrillard responds in terms of his own political life, a life of writing and problematizing. He asks, What organizing principle keeps me at work, if not the driving force of the real? Why write? Isn't writing a cynical exercise if one cannot compel responses? But Baudrillard answers with an exclamation, a performance in the face of shocked surprise that he *could* stop writing unless the real actually intervened.

> For me, it's in the realm of the intellectual wager. If there were an
> absolute term of the nuclear apocalypse in the realm of the real, then at
> that point I would stop, I wouldn't write anymore! God knows, if the
> metaphor really collapses into reality, I won't have any more to do. That
> would not even be a question of resignation — it's no longer possible to
> think at that point.[4]

Baudrillard's conclusion is delightful — he will, of course, go about his business, trying to figure out what politics can be and how all of this image-manufacturing works, and never freak out. Saving that for later, he reminds us, as did Artaud, that true "last reactions" will be crazed and deranged (not liberal, calm, and authoritative). "Then at that point I would stop, I wouldn't write anymore!" The wager is an alternative metaphor we can understand. If the apocalypse were ever real, all bets would be off. Our present context is less dramatic than that; the power of the nuke is soft, unfocused. And if it were otherwise, Baudrillard's (and my own) project would abruptly stop. But that is (merely) obvious and trivial, not nihilistic in the sense intended for poststructuralists by their liberal critics.

Even if no overall agenda is possible, even if our rhetorical claim that all participants in political movement are brothers and sisters becomes untenable, there are still political programs to articulate. As poststructuralism has begun to emerge from its rather theoretical (or perhaps antitheoretical) phase, such political articulations have started to emerge. And this political speculation has found no end of examples in our culture, even if its examples were not taken seriously before these critics embraced them.

When postmodernists write about nukes in terms of 007, *Atomic Cafe*, and crime movie plots, they undermine their own discourse in a way that still hints at an ironic political possibility, even in the face of sobering paradox. This is a political impulse that has already discarded the general, denying the possibility that whole, positive structures could be legitimated by an intellectual class (a Vanguard, to remember the name of one set of U.S. missiles). My discussion of the freeze, for example, does not imply that freezers "thought this way" in establishing their movement, even though several points of their position are represented in their key word.[5] Having long since discarded "program" or political "agenda," but also at a distance from pragmatic opportunism, a postmodern politics might still find ways to oppose domination.

Donna Haraway's essay on cyborgs stands as one of our best examples of the attempt to construct such a political stance. The essay ends with a political statement, suggesting the possibility of a socialist feminism that moves beyond Marxist roots by acknowledging the poststructuralist alteration of nature and origin. The central site of work (and the central model of worker) passes from factory to cottage.

> Work is being defined as both literally female and feminized, whether
> performed by men or women. To be feminized means to be made
> extremely vulnerable; able to be disassembled, reassembled, exploited as
> a reserve labor force.[6]

Insofar as there are new, important factories, they are staffed in industrializing countries by teenage women who are the main sources of cash for their families. Robotics and automated offices intensify this "feminization of work."[7] At the same time, communications technologies undercut public life, creating a "privatization" far more important than anything imagined by neoconservative economists. But Haraway believes a "feminist science/technology politics" could emerge, confronting and rejecting the new forms of technology-enabled power. Even without a "before the fall" mythology to rely on, oppositions can become powerful. And Haraway so enlivens this possibility that her closing line resonates with plausibility, creating the sort of politics poststructuralism requires. "Though both are bound in the spiral dance, I would rather be a cyborg than a goddess."[8]

In Stanley Kubrick's film *Dr. Strangelove*, the end of the world stands for hope.[9] The seemingly inappropriate version of "We'll Meet Again" that accompanies the mushroom cloud conclusion sends news of a break with the past, as does the subtitle, *How I Learned to Stop Worrying and Love the Bomb*. It will be quite a different posture we will have to invent to engage the technology of our times in a politicizing mode. I am arguing that the liberal, humanistic discourse actually resolves questions, allowing us to reify our value choices and avoid politics. In the apocalyptic climax of *Dr. Strangelove*, the computerized plans of all sides are overcome. We have little trouble imagining that, at the end, some of us would not find insight into humanity, but would reenact subservient disciplines, even defending the coins in a soda machine. The heroism of other characters in that film repeatedly works against the human prospect. With a bit of human irrationality, a quandary of strategic moves gone awry, and the pragmatic, typically American, last-minute fix by the archetypal American bomber pilot (portrayed by Slim Pickens), humanity manages to destroy itself. So much for our calm, deliberate attempts to manage technology. Displacing those noble impulses is the ad hoc, impassioned fix by Pickens, who takes the bomb as his body, riding it down with the exuberance of a bronco rider. This works.

When that happens, the questions of whether modern technology is "good" or "bad" are left far behind. Those questions and the deliberate, cautious choice associated with them neither destroy the world nor prevent its destruction. The real engagement of actual persons is, finally, the crux of the matter, even if it is revealed too late, as Pickens exuberantly rides the bomb out of the chute, ending our prospects altogether.

Such an approach is available to us, one hopes to a better end. Once one is engaged with robots, computers, and even the bomb, a conversation is begun with what is best understood as a new agent in the world. We can begin to ask what has been produced in us, how that production has been accomplished, and how the production could be exposed. Activists could begin by understanding how some of our fables have posed a very strange plot. Citizens of the nuclear age are supposed to have been reassured and secured by (were even told to "sleep more soundly" amidst) hugely discomforting technologies that tend to hide or, even, walk autonomously away.

The discourse that would raise those discomforts in a critical manner has hardly begun to be identified, but hints of it emerge from the art world, even from the most popular of art forms. Numerous strategies have been formulated. We could cancel alienation with expressive inarticulateness, or we could array the styles, images, and tropes of the culture against each other, reenacting implosion. To cite just one example from a recently popular rock song, R.E.M.'s apocalyptic ditty rings out with incongruous joyousness, even glee: "It's the end of the world as we know it (and I feel fine)."[10]

These are endless endings, last words that could never be heard, a fall followed by a winter beyond all experience, "an original end of sorts, the final fall of the fall, the spectacular fallout."[11] In the face of such an ending, the nuclear critic can still assemble a package of interventions, rehearsing arguments that will be useful in universities, laboratories, and institutes. It has always been the task of radical political analysis to announce—often joyously—that the "world as we know it" has ended. But this time the task is trickier. Not only the world, but also the most fundamental ways "we know it"—language and discourse—have been undermined.

When the most studious activists come to ask how we are doing these things, somehow beyond the great universalisms they had learned about and relied upon, we could have our arguments (our habits of problematization) ready. My last words: the seemingly irresponsible turns out to have long been appropriate, even unavoidable. The laughing, cynical, fractal, ironic cyborg—at last.

# Notes

## Introduction: Nukes "Я" Us

1. Timothy Egan, "Richland Journal: Little Sentiment Here to Ban the Bomb," *New York Times*, January 14, 1988, sec. A, 14.

2. "Students Back N-Mushroom Cloud," *Spokane Spokesman-Review*, February 24, 1988, sec. A, 1.

3. Ibid.

## 1. Knowing Nukes

1. Race, the third world, and gender have received particular attention. See Henry Louis Gates, Jr., ed., *"Race," Writing, and Difference* (Chicago: University of Chicago Press, 1986); Nancy Hartsock, "Rethinking Modernism: Minority vs. Majority Theories," *Cultural Critique* 7 (Fall 1987): 187-206; and Edward Said, *The World, the Text, and the Critic* (Cambridge, Mass.: Harvard University Press, 1983) for two examples, though the field is much more crowded than any listing here could suggest.

2. Jonathan Schell, *The Fate of the Earth* (New York: Knopf, 1982).

3. See Robert Jay Lifton, *The Future of Immortality* (New York: Basic Books, 1987), 111-35.

4. Jonathan Schell, *The Abolition* (New York: Knopf, 1984).

5. George Kateb, "Thinking about Human Extinction (I): Nietzsche and Heidegger," *Raritan* 6 (Fall 1986): 1-28, and "Thinking about Human Extinction (II): Emerson and Whitman," *Raritan* 6 (Winter 1987): 1-22.

6. Kateb, "Thinking about Human Extinction (I)," 7.

7. Ibid., 16-17.

8. Ibid., 7-8.

9. Kateb discusses this act of establishing value internal to humanity in terms of the attempt to reach Arendt's Archimedean point, or redefine that epistemological ambition if it

is, indeed, impossible for us to be the judge in our own case. "Thinking about Human Extinction (I)," 13.

10. Ibid., 2-3.

11. Ibid., 3-4.

12. Kateb takes this move even further, extrapolating from this individualism a political mode, a way of acting in contemporary settings where action and reflection have been unlikely twins: "The player tends to adopt a spectatorial perspective on his own activity as well as that of others. When at rest, when engaged in contemplation or rumination, the observer feels with, sympathizes or empathizes with, those who act and suffer. Thus action allows itself contemplative moments, and contemplation is ravenously active." Kateb, "Thinking about Human Extinction (II)," 18.

13. Ibid.

14. Ibid.

15. Ibid., 22.

16. Michel Foucault, *Power/Knowledge: Selected Interviews and Other Writings, 1972–1977*, ed. Colin Gordon (New York: Pantheon, 1980), 126.

17. Ibid., 127.

18. Ibid., 127-8. Elsewhere, Foucault went further, identifying the post-Hiroshima era as the actual beginning of the contemporary age, so significant that there were only two other eras of modernity.

19. Barry Cooper, "The Meaning of Technology at the End of History," *University of Minnesota Center for Humanistic Studies Occasional Papers* 7 (1986): 23. The *Wibakusha* quote appeared in Robert Jay Lifton, *Death in Life: Survivors of Hiroshima* (New York: Random House, 1967), 79. Lifton is quoted from Robert Jay Lifton and Eric Olson, *Living and Dying* (New York: Praeger, 1974), 23.

20. Jacques Derrida, "No Apocalypse, Not Now (full speed ahead, seven missiles, seven missives)," *Diacritics* 14 (Summer 1984): 20-31. The Cornell University symposium that eventually produced the *Diacritics* issue is discussed at length in J. Fisher Solomon, *Discourse and Reference in the Nuclear Age* (Norman: University of Oklahoma Press, 1988), esp. 5-21. Solomon (17) reads Derrida's essay as "a challenge to anyone seeking to cross unambiguously from the critical text into an extracritical reality," a contribution that "undercut from the start the very project that the conference had been intended to inaugurate." Accordingly, Solomon deals with the essay at much more length than have I.

21. This use of "textuality" may require some introduction for those not familiar with recent literary criticism. A full review is beyond the scope of this book, but, briefly, the claim is that the most appropriate way of approaching the political world is as if it were text or texts, accessible to the sort of criticism one does on literary texts. This contrasts with the scientific view, which finds data in the world, for example. For more on the applicability of "textualism" in a political science context, see Michael Shapiro, *Language and Political Understanding* (New Haven: Yale University Press, 1981), and Michael Shapiro, ed., *Language and Politics* (New York: New York University Press, 1984).

22. Derrida, "No Apocalypse," 23. See also Michael S. Sherry, *The Rise of American Air Power: The Creation of Armageddon* (New Haven: Yale University Press, 1988). In a review, Garry Wills wrote: "That dissociation of destruction from victory is the distinguishing mark of modern war, and Sherry shows that, throughout World War II, the technological procedures of *destroying* failed, over and over, to mesh with strategic evaluation of the means toward *prevailing*. The two were not brought into accord because people assumed they were inseparable." Garry Wills, "Aerie Visions," *New York Review of Books*, March 17, 1988, 7.

23. Derrida, "No Apocalypse," 23. In deconstructionist usage, the break corresponds to the break between "language" and "reality"; i.e., to *aporia*.

24. Ibid.

25. Peter Schwenger, "Writing the Unthinkable," *Critical Inquiry* 13 (Autumn 1986): 33-48.

26. The primary point of Schwenger's article is to address this possibility, giving examples of literary criticism of modern novels that center on nukes.

27. This question of visibility is the topic of chapter 3, where I will look at several examples as well as the contrary argument that militarism has a pervasive presence, not an invisible one.

28. Frances Ferguson, "The Nuclear Sublime," *Diacritics* 14 (Summer 1984): 4-11.

29. Solomon, *Discourse and Reference*, 45.

30. Ibid., 46, 61, 73.

31. Ibid., 68.

32. Ibid., 207-11, 263-69.

33. Ibid., 111.

34. Michael McCanles, "Machiavelli and the Paradoxes of Deterrence," *Diacritics* 14 (Summer 1984): 14.

35. Ibid., 19.

36. Richard Klein and William B. Warner, "Nuclear Coincidence and the Korean Airline Disaster," *Diacritics* 16 (Spring 1986): 2-21.

37. Ibid., 5.

38. Ibid., 6.

39. The full quotation plays with the familiarity with "spy genre" thrillers; these are human foibles we understand very well, in the literary (if not the political) world. If we do imagine that this was a spy mission (we don't know, with any certainty, either way), we can imagine that "it might have never occurred to anyone to make the connection between the written and the oral form of the number, until the moment when the number came screaming over the news wires, and the Company, as at a bad pun, let out a collective groan. The groan is a sign of embarrassment, the discomfort when something that should have remained hidden is revealed. One can suppose that 007 was an instance of what the spy most fears—the recurrent nightmare—the capacity of his unconscious to betray him into letting slip what he has taken every conceivable rational precaution to conceal." Klein and Warner, "Nuclear Coincidence," 8.

40. Ibid., 5.

41. Garry Wills, "Critical Inquiry (*Kritik*) in Clausewitz," *Critical Inquiry* 9 (December 1982): 287.

42. Ibid.

43. For a discussion of danger and how it has transformed in the contemporary era, see Thomas Dumm, *Democracy and Punishment: Disciplinary Origins of the United States* (Madison: University of Wisconsin Press, 1987).

44. Wills, "Critical Inquiry," 297-98.

45. Klein and Warner, "Nuclear Coincidence," 8.

46. Dean MacCannell, "Baltimore in the Morning . . . After: On the Forms of Post-Nuclear Leadership," *Diacritics* 14 (Summer 1984): 34-46. The Einstein quotation originally appeared in Albert Einstein, "The Real Problem Is in the Hearts of Men," *New York Times Magazine*, June 23, 1946, 43.

47. Jean Baudrillard, "Forget Baudrillard: An Interview with Sylvere Lotringer," in *Forget Foucault* (New York: Semiotext[e], 1987), 109.

48. Michel Foucault, *The History of Sexuality: Volume I: An Introduction* (New York: Vintage, 1978), and *Discipline and Punish: The Birth of the Prison* (New York: Vintage, 1979).

49. See Chellis Glendinning, *Waking Up in the Nuclear Age: The Book of Nuclear Therapy* (New York: William Morrow, 1987). In fact, the author tries to find political consequences within therapeutic language. A poststructuralist reading would surely note the ability of therapy to confound such political intentions. A clearer example of the therapeutic politics of nukes was reported in "DOE Grants $85,000 to Halt 'Nuclear Phobia.' " *Not Man Apart* (December 1984): 15.

50. Jean Baudrillard, *Fatal Strategies*, trans. Philip Beitchman and W. G. J. Niesluchowski, ed. Jim Fleming (New York: Semiotext[e], 1990), 17.

51. Edward Said, *The World, the Text, and the Critic* (Cambridge, Mass.: Harvard University Press, 1983), 4.

52. Ibid., 174. See Said's comment on the disagreements between Derrida and Foucault, in "Criticism Between Culture and System," ibid., 178-225.

53. See Stephen Hilgartner, Richard C. Bell, and Rory O'Connor, *Nukespeak: Nuclear Language, Visions, and Mindset* (San Francisco: Sierra Club Books, 1982).

54. Herbert Marcuse, *One-Dimensional Man: Studies in the Ideology of Advanced Industrial Society* (Boston: Beacon Press, 1964), 93.

## 2. No More Warriors

1. The method of change—the mass, televised demonstration—is another story, which I will discuss elsewhere.

2. Just as obviously, this chapter was completed well before the Gulf War (and its Kurdish aftermath) of 1991. I believe the analysis survived the war, but I have chosen not to alter it significantly. In a speeding world, the author needs to stop somewhere.

3. And, yes, now sometimes she. I will use the male pronoun, in deference to the claim that his existence is crucially male.

4. Michel Foucault, *Discipline and Punish: The Birth of the Prison* (New York: Vintage, 1979), 135.

5. Ibid.

6. Ibid., 169.

7. Gilles Deleuze and Félix Guattari, *Nomadology: The War Machine* (New York: Semiotext[e], 1986), 13.

8. Ibid., 110.

9. A stubborn reluctance to recognize this development has made the 1990 debate about the fate of Germany an odd and nostalgic enterprise. Somehow, the turf of Germany remains as important as it ever was, despite all the technological developments of the last half century.

10. Paul Virilio, *Speed and Politics: An Essay on Dromology* (New York: Semiotext[e], 1986), 56.

11. This question of velocity is also crucial to Derrida's treatment of nukes, discussed elsewhere in this book. See Jacques Derrida, "No Apocalypse, Not Now (full speed ahead, seven missiles, seven missives)," *Diacritics* 14 (Summer 1984): 20-31.

12. Peter Sloterdijk, *Critique of Cynical Reason* (Minneapolis: University of Minnesota Press, 1987), 129.

13. Ibid., 353-54, 448.

14. Avital Ronell, "Starting from Scratch: Mastermix," *Socialist Review* 18 (April-June, 1988): 77.

15. Paul Virilio, *War and Cinema: The Logistics of Perception* (London: Verso, 1989), 83.

16. Seymour M. Hersh, *"The Target Is Destroyed": What Really Happened to Flight 007 and What America Knew About It* (New York: Random House, 1986), 121.

17. Michael McCanles, "Machiavelli and the Paradoxes of Deterrence," *Diacritics* 14 (Summer 1984): 15-24.

18. Ibid., 17-18. The more widely publicized version of this statement was part of a leaked National Security Council document of the early 1980s. That document spoke of a U.S. "policy to prevail in a protracted nuclear war." Its publication drew a rebuttal from Secretary of Defense Casper Weinberger. For a discussion of the historical context, see Theodore Draper, *Present History: On Nuclear War, Détente, and Other Controversies* (New York: Random House, 1983). An extensive and useful analysis of Weinberger's response is presented in J. Fisher Solomon, *Discourse and Reference in the Nuclear Age* (Norman: University of Oklahoma Press, 1988), 151-54.

19. Timothy W. Luke, *Screens of Power: Ideology, Domination, and Resistance in Informational Society* (Urbana: University of Illinois Press, 1989), 90.

20. Of course, it had also suggested how one might delegitimize that "agreement"; the drafting of young — often poor and black — men into the Marines was yet another Vietnam-era shock.

21. Garry Wills, "Critical Inquiry (*Kritik*) in Clausewitz," *Critical Inquiry* 9 (December 1982): 283.

22. This connection was suggested to me by Doug Borer. The "Maytag repairman" stands in for all sorts of contemporary roles. He is, we should note, an actor who has fashioned a whole career, portraying an employee who is notable only for never having anything to do.

23. This issue of "danger" is itself an important topic of nuclear criticism. See Thomas Dumm, *Democracy and Punishment: Disciplinary Origins of the United States* (Madison: University of Wisconsin Press, 1987).

24. Jean Baudrillard, *Simulations*, trans. Paul Foss, Paul Patton, and Philip Beitchman (New York: Semiotext[e], 1983), 32.

25. "Transgression and violence are less serious, for they only contest the distribution of the real. Simulation is infinitely more dangerous, however, since it always suggests, over and above its object, that law and order themselves might really be nothing more than a simulation." Baudrillard, *Simulations*, 38.

26. Ibid., 41.

27. Ibid., 59-60.

28. Ibid., 63, 64.

29. Ibid., 68.

30. Ibid., 70.

31. Ibid., 73. What does lie outside that realm of control is terrorism, but it, too, has been deprived of content. For Baudrillard, the terrorist stands at the culmination of the postmodern warrior, beyond either "banditry" or "commando action." (These are the primary charges levied against Oliver North, we should note.) Without "economic or martial" logics we have known — no longer representing (legitimacy claims or historical continuities) — terrorism becomes the classic military form of our era. "Its blindness is the exact replica of the system's lack of differentiation." Jean Baudrillard, *In the Shadow of the Silent Majorities . . . Or the End of the Social: And Other Essays*, trans. Paul Foss, John Johnston, and Paul Patton (New York: Semiotext[e], 1983), 55-56.

32. See Michel Foucault, *Power/Knowledge: Selected Interviews and Other Writings, 1972-1977*, ed. Colin Gordon (New York: Pantheon Books, 1980), 127-28.

33. Baudrillard, *Simulations*, 59-60.

34. Ibid., 73.

35. Quoted by Ellen Goodman, "The Summit's Key Value: Openness to Change," *Reno Gazette-Journal*, December 16, 1987, sec. A, 17.

36. Fawn Hall was the spy's spy, a true poststructuralist actor who took the purloined letters and attached them to her body in a futile attempt to make them invisible.

37. James Der Derian, *On Diplomacy: A Genealogy of Western Estrangement* (New York: Basil Blackwell, 1987), 205-6, his emphasis.

38. Ibid., 206.

39. Ibid., his emphasis.

40. Ibid., 207-8.

41. Michel Foucault, *The History of Sexuality: Volume I: An Introduction*, trans. Robert Hurley (New York: Vintage, 1978).

42. For readers familiar with American political thought, yet one more reading of the leukemia metaphor might focus on the moral value of politics, a central concept for democratic theory. Especially since Dewey, liberal pragmatists have bemoaned a loss (or thinness) of political life. They perceived an active, social consideration of power as valuable almost regardless of its (material or institutional) content. It was virtually enough, to fuel the pragmatist's critique, that the opportunities for this kind of activity were diminishing. When critical theory turned to humanist readings of Marx, this loss of politics was expressed as an aspect of alienation. In any case, the demise of politics—an internal loss of ability to perform, regardless of external context—would match Baudrillard's leukemization.

43. See Jean Baudrillard, *Forget Foucault* (New York: Semiotext[e], 1987).

44. Jean Baudrillard, *Simulacres et simulation* (Paris, Editions Galilée, 1981), 229. Quoted in and translated by Paul Foss, "Despero Ergo Sum," in *Seduced and Abandoned: The Baudrillard Scene*, ed. André Frankovits (New York: Semiotext[e], 1984), 11-12.

45. Foss, "Despero Ergo Sum," 14-15.

46. This is all the more attractive a term, now that it is a therapeutic term of art; drug counselling clinics now offer "intervention" as a technique that can help "even those who don't think they need help."

## 3.  Robotics (The Bomb's Body)

1. Andreas Huyssen, "The Vamp and the Machine: Technology and Sexuality in Fritz Lang's Metropolis," *New German Critique* 24-25 (Fall-Winter 1981–82): 221-37.

2. Langdon Winner, *Autonomous Technology: Technics-out-of-Control as a Theme in Political Thought* (Cambridge: MIT Press, 1977), esp. chapters 1 and 2.

3. Donna Haraway, "A Manifesto for Cyborgs: Science, Technology, and Socialist Feminism in the 1980s," *Socialist Review* 15 (March-April 1985): 64-107.

4. Ibid., 66.

5. Ibid., 67-70.

6. Our ambivalence toward arms control—which has hinged on "verification"—may mark this transition. After all, what are we to make of agreements that assume a visibility we have learned so well to live without? During the maneuvering between Gorbachev and Reagan in the summer of 1987, this ambivalence was played out on the stage of diplomacy, as well as in conservative domestic politics.

7. When the trains came to be known colloquially as "white trains," their owners painted the trains—which were of an unmistakable shape—multiple colors. In doing so, they were entering into a ritual of the visible and invisible, reinforcing the importance of visibility, even if they fooled nobody.

8. Ellen Willis, *Village Voice*, April 30, 1979, 8.

9. Ibid. Willis continues: "And after all, since Hitler's genocide was an atrocity unprecedented in history, who could have conceived of it, or anything close to it, beforehand? Even those who had intimations of how bad the situation could become might easily have

dismissed them as irrational panic. I have no idea what I would have done. All I know is, New York is 30 miles from a nuclear plant on a fault line, and here I am."

10. See, for example, the widely read predictions of the author of BASIC, the elementary programming language used by many beginners. John G. Kemeny, *Man and the Computer* (New York: Scribner's, 1972).

11. These efforts, an early and characteristically sixties Berkeley intervention, are detailed in Steven Levy, *Hackers: Heroes of the Computer Revolution* (Garden City, N.Y.: Doubleday, 1984). Felsenstein would later moderate the Homebrew Computer Club in Palo Alto (the point of emergence of the first Apple computer), and later yet would design the popular Osborne microcomputer.

12. Levy's *Hackers* contains a definitive discussion of the hacker ethic.

13. Seymour Papert, *Mindstorms: Children, Computers, and Powerful Ideas* (New York: Basic, 1980). The name LOGO is a kind of bastardized pun, the kind of singular of "logos" only a computer would generate. Hence, it proposes the further distillation of knowledge (logos), a procedure the Greeks would have quickly dismissed. LOGO is, thus, a contemptuous honorific.

14. Minsky's more recent ruminations are marked by sophistication and, even, caution. See Marvin Minsky, *The Society of Mind* (New York: Simon & Schuster, 1986). For a discussion of his earlier utopian urge, see Hubert Dreyfus, *What Computers Can't Do*, rev. ed. (New York: Harper & Row, 1979), esp. 80-81. A somewhat more appealing utopianism (because it is more cantankerous and, even, political) was circulated by Ted Nelson as early as 1974. It has been reprinted and revised, in Ted Nelson, *Computer Lib* (Redmond, Wash.: Tempus Books, 1987).

15. Sherry Turkle, *The Second Self: Computers and the Human Spirit* (New York: Simon & Schuster, 1984), 29-63.

16. Haraway, "A Manifesto," 70-71.

17. Ibid., 70.

18. Dreyfus, *What Computers Can't Do*, and John R. Searle, "Minds, Brains, and Programs," in *The Mind's I*, ed. Douglas R. Hofstadter and Daniel C. Dennett (New York: Bantam Books, 1981), 353-82.

19. Winner, *Autonomous Technology*, 54. See also Arthur O. Lewis, Jr., ed., *Of Men and Machines* (New York: E. P. Dutton, 1963), 198-204.

20. Huyssen notes that Frankenstein was the first instance of the turn toward a dangerous, fearsome robotic technology. See also the discussion of Frankenstein in Winner, *Autonomous Technology*, 306-17.

21. Jean Baudrillard, *For a Critique of the Political Economy of the Sign*, trans. Charles Levin (St. Louis: Telos Press, 1981), 86. See also Baudrillard, *Simulations*, trans. Paul Foss, Paul Patton, and Philip Beitchman (New York: Semiotext[e], 1983), 92-96.

22. The "Asimov Priorities," as they are called in science fiction, are (in order of importance): protect humans, obey humans, and protect yourself.

23. See Dreyfus, *What Computers Can't Do*, for a detailed discussion of AI hubris. An interesting comparison is found in Stewart Brand, *The Media Lab: Inventing the Future at MIT* (New York: Penguin, 1988). Visiting a different part of MIT, over two decades after the early days of AI, Brand finds an organization more savvy about budgets and time lines, but one still driven by the logic of public relations. The old academic "publish or perish" had been displaced by a postmodern version; "demo [as in "demonstrate" a product or approach] or die."

24. "All Things Considered," National Public Radio, August 27, 1987.

25. Seymour M. Hersh, *"The Target Is Destroyed": What Really Happened to Flight 007 and What America Knew About It* (New York: Random House, 1986), 133.

26. Diane Rubenstein, "Hate Boat: Greenpeace, National Identity, and Nuclear Criticism," in *International/Intertextual Relations: Postmodern Readings of World Politics*, ed. James Der Derian and Michael J. Shapiro (Lexington, Mass.: Lexington Books, 1989), 231-56.

27. See Turkle, 201, for a discussion of Pirsig's hacker fans. The AI community's enthusiasm for Robert Pirsig's *Zen and the Art of Motorcycle Maintenance: An Inquiry Into Value* (New York: William Morrow, 1974) bespeaks a commitment to the computer as a setting for ontological and metaphysical search. This, finally, is where we can find out what human beings are. Although Pirsig's ostensible project is a spiritual anti-Luddite ethos, he implies this broader agenda when he disputes the humanist notion that technologies are always "mere tools," instrumental to our intended tasks. In short, Pirsig urges his readers to respond to the great puzzles of computers by moving to find new metaphysical categories. Pirsig argues that certain technologies—the ones a person most cares about—exceed the limits implied by their status as "things," becoming more like autonomous agents that can participate in the most fundamental questioning a person can do.

28. See the discussion of metonymy in Diane Rubenstein, "Food for Thought: Metonymy in the Late Foucault," in *The Final Foucault*, ed. James Bernauer and David Rasmussen (Cambridge: MIT Press, 1988), 83-101.

29. Jacques Lacan, *The Structuralists from Marx to Levi-Strauss* (Garden City, N.Y.: Anchor, 1972), 310.

30. Gilles Deleuze and Félix Guattari, *Anti-Oedipus: Capitalism and Schizophrenia*, trans. Robert Hurley, Mark Seem, and Helen R. Lane (Minneapolis: University of Minnesota Press, 1983), 114-15.

31. The enthusiasm of this dance is chronicled in Suzanne Campbell-Jones (writer and producer), "Talking Turtle" (London: BBC TV, 1983, also shown on PBS, 1983).

32. Deleuze and Guattari, *Anti-Oedipus*, 8.

33. Dean MacCannell, "Baltimore in the Morning . . . After: On the Forms of Post-Nuclear Leadership," *Diacritics* 14 (Summer 1984): 34.

34. Ibid.

35. The phrase refers to the subtitle of Kubrick's *Dr. Strangelove, or How I Learned to Stop Worrying and Love the Bomb* (Great Britain: Hawk Films, 1963). It is remarkable how many spectacles that movie managed to predate, as well as to chronicle. Dan Baum has pointed out to me that Slim Pickens's inventory of the emergency kit originally evoked a comment from him about Dallas, not Las Vegas. The dub is visible on screen; Pickens' lips say "You could have a pretty good time in Dallas with all this stuff." The reason for the dub is visible in the film's release date.

36. MacCannell's ("Baltimore," 42) provocative, selective list (arranged so that it mixes the seemingly ridiculous and seemingly sublime) includes: explicit nuclear references in Heavy Metal music; the love affair with four-wheel-drive vehicles; the seemingly anachronistic popularity of country music; our Western sheriff presidency; the officially supported renewal of fatalistic Protestantism; emphasis on highway rather than rail transport; development of alternate (wind, solar) energy sources; the breakup of AT&T; the rapid diffusion of the home computer; and, even, the popularity of jogging as a form of exercise. The cultural context of nuclearism is the topic of another article in the nuclear criticism symposium. See Derrick De Kerckhove, "On Nuclear Communication," *Diacritics* 14 (Summer 1984): 72-81.

37. To a degree, nuclear opponents have been claiming that nuclearism is a culture, a point nuclear criticism would broaden. See Paul Rogat Loeb, *Nuclear Culture* (New York: Coward, McCann & Geoghegan, 1982).

38. MacCannell, "Baltimore," 45.

39. *Desert Bloom*" (Los Angeles: Columbia Pictures, 1986). This contemporary classic was produced by Robert Redford as part of the Sundance Institute's project for developing

independent American films. Hiding their thoroughly postmodernist presumptions behind a facade of slick Hollywood filmmaking, director-author Eugene Corr and coauthor Linda Remy may have hidden too well; the film was seldom shown.

40. Deleuze and Guattari, *Anti-Oedipus*, 335.

41. Arthur Kroker and David Cook, *The Postmodern Scene: Excremental Culture and Hyper-Aesthetics* (New York: St. Martin's Press, 1986), 185-86.

42. Ibid.

43. Huyssen, "The Vamp and the Machine," 221-37.

44. Foucault made this clear in his reaction to Mitterrand's election. See "Practicing Criticism," in Michel Foucault, *Politics, Philosophy, Culture: Interviews and Other Writings of Michel Foucault, 1977-1984,* ed. Lawrence D. Kritzman (New York: Routledge, 1988), 152-56.

45. Birrell Walsh, "The Monkey Trap or the Mystical Engine?" *Whole Earth Review* 44 (January 1985): 1-2. Walsh was among the inventors of the Apple Macintosh computer.

46. Walsh, "The Monkey Trap," 2.

## 4.  Star Wars and the Freeze

1. Diane Rubenstein, "The Mirror of Reproduction: Baudrillard and Reagan's America," *Political Theory* 17 (November 1989): 599, 600, 604.

2. See Michael Rogin, *"Ronald Reagan," the Movie* (Berkeley: University of California Press, 1987). Rogin places Reagan's approach within a broad tradition of demonology and countersubversion.

3. For more conventional reports on the freeze as a policy approach and as evidence of political movement, see Douglas C. Waller, *Congress and the Nuclear Freeze* (Amherst: University of Massachusetts Press, 1987), and Paul Rogat Loeb, *Hope in Hard Times: America's Peace Movement and the Reagan Era* (Lexington, Mass.: D.C. Heath, 1987).

4. For another poststructuralist treatment of the nuclear freeze, see Avital Ronell, "Starting from Scratch: Mastermix," *Socialist Review* 18 (April-June 1988): 74-85.

5. Physicist Alvin Weinberg is quoted in Marshall Berman, *All That Is Solid Melts into Air: The Experience of Modernity* (New York: Simon & Schuster, 1982), 84-85.

6. Jacques Derrida, "No Apocalypse, Not Now (full speed ahead, seven missiles, seven missives)," *Diacritics* 14 (Summer 1984): 20-31, and Paul Virilio, *Speed and Politics* (New York: Semiotext[e], 1986).

7. Virilio, *Speed and Politics*, 37.

8. Timothy O'Brien, *The Nuclear Age* (New York: Knopf, 1985).

9. Martin Robertson and Alison Frantz, *The Parthenon Frieze* (New York: Oxford University Press, 1975).

10. Jean Baudrillard, *Simulations*, trans. Paul Foss, Paul Patton, and Philip Beitchman (New York: Semiotext[e], 1983), 61, 74.

11. George Lipsitz's analysis of Reagan centers on these heroes, showing that Reagan's speeches "constructed the audience as historical subjects with a common past." Lipsitz contrasts this construction with another—Jesse Jackson's more diverse project. Both are constructions, not simply histories. See George Lipsitz, *Time Passages: Collective Memory and American Popular Culture* (Minneapolis: University of Minnesota Press, 1990), 32-34.

12. The themes of Western, cowboy mythology, and defense policy are discussed in Garry Wills, *Reagan's America: Innocents at Home* (Garden City, N.Y.: Doubleday, 1987). See also Michael Rogin, *"Ronald Reagan," the Movie*. As Wills explains, Reagan's "lone marshall" mythology was even unfaithful to its origins. Most often, in the American West, order was reimposed by communities and civil authorities, acting through the same mechanisms of state coercion that prevailed throughout Europe and the eastern U.S. at the time.

The true civilizers of the frontier were not lone gunmen, "tall in the saddle" in a favorite Reagan phrase, romantic in their heroic search for good, but agents of civil society — troops of town marshals, gun controllers, and moralizers from the ubiquitous churches.

13. Michael McCanles, "Machiavelli and the Paradoxes of Deterrence," *Diacritics* 14 (Summer 1984): 19.

14. Virilio, *Speed and Politics*, 143.

15. Andy Pasztor and Bob Davis, "Battle Plans: If War Comes, the U.S. Hopes to Make It Brief With High-Tech Blitz," *Wall Street Journal*, September 6, 1990, sec. A, 1, 6.

16. Reagan's quirky perception of nuclear weapons has been widely chronicled, perhaps best by Strobe Talbott, *Deadly Gambits: The Reagan Administration and the Stalemate in Nuclear Arms Control* (New York: Knopf, 1984).

17. Garry Wills, "Critical Inquiry (*Kritik*) in Clausewitz," *Critical Inquiry* 9 (December 1982), 284.

18. George Lucas (writer and director), *Star Wars* (Los Angeles: Twentieth Century-Fox, 1977).

19. In the second part of the series, *The Empire Strikes Back* (Los Angeles: LucasFilm, 1980), written by George Lucas, the stand-in for the child is the organic Yoda, and the theme is faith and power. Released in the year of Reagan's election to president, the ideological similarity was close enough to assure the series' influence. But the patterns were set and the later sequel became less interesting.

20. Fredric Jameson, "Postmodernism or the Cultural Logic of Late Capitalism," *New Left Review* 146 (July-August 1984): 66-67. Rubenstein, "The Mirror of Reproduction," applies this idea to Ronald Reagan as well: "For Ronald Reagan, the real comes back as a double of a self that never was" (583).

21. See Paul Virilio, *Speed and Politics*, 137-42. Virilio's prediction (first published in 1977, before SDI became a major policy issue) argued that modern computer control capabilities would surely end up helping to abolish politics.

22. For a discussion of literary production and politics, see Michael J. Shapiro, *The Politics of Representation: Writing Practices in Biography, Photography, and Policy Analyses* (Madison: University of Wisconsin Press, 1988), 3-54, and "Representing World Politics: The Sport/War Intertext," in *International/Intertextual Relations: Postmodern Readings of World Politics*, ed. James Der Derian and Michael J. Shapiro (Lexington, Mass.: Lexington Books, 1989), 71: "It is important to recognize that to employ a textualizing approach to social policy involving conflict and war is not to attempt to reduce social phenomena to various concrete manifestations of language. Rather, it is an attempt to analyze the interpretations governing policy thinking. And it is important to recognize that policy thinking is not unsituated. While it is expressed in various dominant forms of representation, those representational practices arise out of a society's more general practices (for example, the modes through which various social spaces are produced). Therefore, this analysis seeks both to discern the representational practices that construct the "world" of persons, places, and modes of conduct and to inquire into the network of social practices that give particular modes of representation their standing."

23. For an extended reading of Machiavelli from a language-and-politics position, see R. B. J. Walker, "*The Prince* and 'The Pauper': Tradition, Modernity, and Practice in the Theory of International Relations," in Der Derian and Shapiro, *International/Intertextual Relations*, 25-48. Walker contests the view that Machiavelli was the "archrealist," suggesting instead that his topic was "political life within states" (34). "More than almost anyone else, certainly more than either Hobbes or Rousseau, it is Machiavelli who symbolizes what the tradition of international relations theory is all about. . . . To take Machiavelli seriously is to confront . . . someone who can be read in ways that problematize the most basic assumptions on which claims about the tradition are based. Contrary to both the so-called

realists who treat Machiavelli as one of their own and the so-called idealists who castigate him for his supposed realism, Machiavelli poses questions about political community and practice" (29).

24. Berman, *All That Is Solid*, 94.

25. Quoted from Robert C. Tucker, ed., *The Marx-Engels Reader*, 2d edition (New York: Norton, 1978), 338, in Berman, *All That Is Solid*, 95.

## 5.  Immodest Modesty

1. Michel Foucault, "Truth and Power" (interview conducted and translated by A. Fontana and P. Pasquino), in *Power/Knowledge: Selected Interviews and Other Writings, 1972-1977*, ed. Colin Gordon (New York: Pantheon, 1980), 126.

2. R. B. J. Walker, "Genealogy, Geopolitics and Political Community: Richard K. Ashley and the Critical Social Theory of International Politics," *Alternatives* 13 (January 1988): 86.

3. In one important example, the substantial discussion in *International Studies Quarterly* has included M. A. Maggiotto and E. R. Wittkopf, "American Public Attitudes Toward Foreign Policy," *International Studies Quarterly* 25 (December 1981): 601-31; Thomas Ferguson, "The Right Consensus? Holsti and Rosenau's New Foreign Policy Belief Surveys," *International Studies Quarterly* 30 (December 1986): 411-23; C. W. Kegley, Jr., "Assumptions and Dilemmas in the Study of Americans' Foreign Policy Beliefs: A Caveat," *International Studies Quarterly* 30 (December 1986): 447-71; O. R. Holsti and J. N. Rosenau, "Consensus Lost. Consensus Regained? Foreign Policy Beliefs of American Leaders, 1976-1980," *International Studies Quarterly* 30 (December 1986): 375-409; O. R. Holsti and J. N. Rosenau, "The Foreign Policy Beliefs of American Leaders: Some Further Thoughts on Theory and Method," *International Studies Quarterly* 30 (December 1986): 473-84; E. R. Wittkopf, "On the Foreign Policy Beliefs of the American People: A Critique and Some Evidence," *International Studies Quarterly* 30 (December 1986): 425-45; and E. R. Wittkopf, "Elites and Masses: Another Look at Attitudes Toward America's World Role," *International Studies Quarterly* 31 (June 1987): 131-59.

4. For further background on the approach I am invoking, see Michael J. Shapiro, "Textualizing Global Politics," in *International/Intertextual Relations: Postmodern Readings of World Politics*, ed. James Der Derian and Michael J. Shapiro (Lexington, Mass.: Lexington Books, 1989), 9-22. Shapiro suggests how this displacement might also be turned back onto political science itself: "Textually oriented approaches apply discursive rather than psychological analysis and produce highly politicized modes of understanding. . . . Rather than measuring attitudes to test propositions about what social and psychological characteristics account for various individual attitudes, political and otherwise, a discourse approach would encourage raising the question of how it is that the phenomenon of the "attitude" found its way into the speech and writing practices and analyses of the political science profession. (Indeed, it is one of the most highly funded 'objects' of attention in the discipline)" (16).

5. Wittkopf, "Elites and Masses," 133.

6. Holsti and Rosenau, "Consensus Lost. Consensus Regained?" 375, and "The Foreign Policy Beliefs of American Leaders," 473.

7. Maggiotto and Wittkopf, "American Public Attitudes Toward Foreign Policy," 614, and Wittkopf, "On the Foreign Policy Beliefs of the American People," 425.

8. Ferguson, "The Right Consensus?" 417, 415.

9. Or, for that matter, engaging the substantial debate over this method that has already begun. See Kegley, "Assumptions and Dilemmas," and the response by Holsti and Rosenau, "The Foreign Policy Beliefs of American Leaders."

10. Perhaps the most prominent expression of "TGAL" is in René Dubos, *Celebrations of Life* (New York: McGraw, 1981).

11. Seery also uses bumper stickers as an illustration of the ironic in this discourse. John Evan Seery, *Political Returns: Irony in Politics and Theory from Plato to the Antinuclear Movement* (Boulder, Colo.: Westview Press, 1990) "The anti-nuclear bumpersticker, 'One Nuclear Bomb Will Ruin Your Entire Day,' echoes and plays upon another prevalent bumpersticker theme, 'Have a Nice Day!' The anti-nuclear version imitates the facile tone of the 'Nice Day' bumpersticker; but it also parodies and thus subverts the ostensibly affable tone of this genre by juxtaposing the threat of catastrophe with the concern for daily niceness" (305-7).

12. This is only one of several exclusions that could be traced from out of the shadows of the elite opinion model. Two other examples deserve mention. In one case, straightforward pleas by constituents of specific diverse interests (American Jews or Hispanics concerned about Central America) are combined into an "internationalist" position that is secondary, in their cases, to an interest in particular struggles against specific grids of power. Secondly, there is the case of explicitly intermittent claims—the irregular, discontinuous, fleeting outrages that organize around a moment, then disappear to rebel another day. The students of Camus who protested U.S. entry into Cambodia, but did not form an ongoing, ideological cover for their protest, are seen in this debate as a "failed reformulation," not as a newly emergent aspect of politics in the form of the rebel and the spectacle, as poststructuralists might well see them. See Albert Camus, *The Rebel: An Essay on Man in Revolt* (New York: Knopf, 1957).

13. An analysis of the European movements would be, I think, entirely different from this one. Their presence in (at least some) electoral scenes alters their tactics, in turn altering their movement in the most fundamental ways. The values they share with U.S. Greens diminish as a basis for transatlantic unity, which instead forms over collective images and signs.

14. Richard K. Ashley, "Geopolitics, Supplementary, Criticism: A Reply to Professors Roy and Walker," *Alternatives* 13 (January 1988): 91.

15. Jean Baudrillard, *In the Shadow of the Silent Majorities . . . Or the End of the Social*, trans. Paul Foss, John Johnston, and Paul Patton (New York: Semiotext[e], 1983), 68.

16. See the translator's note (Baudrillard, *Silent Majorities*, 1). I am using the term "masses" in the context of Baudrillard's argument. Many American activists now find this term highly disturbing, enacting their own poststructuralist confirmation of individuation. The "sense of self worth" they insist upon is at the core of my argument.

17. Baudrillard, *Silent Majorities*, 6-7.

18. Ibid., 14.

19. Ibid., 16.

20. Ibid., 17, 18.

21. Jean Baudrillard, *For a Critique of the Political Economy of the Sign* (St. Louis: Telos Press, 1981), 147-48.

22. Baudrillard, *Silent Majorities*, 22-23.

23. Ibid., 58.

24. This connection was suggested to me by Fred Rice.

25. For a discussion of apolitical tendencies in Foucault's position, see Edward Said, *The World, the Text, and the Critic* (Cambridge, Mass.: Harvard University Press, 1983), 182-90, 219-22, 244-47. Said concedes the rules-formation process identified by Foucault, but insists there is a further, "historical" process at work: "[Foucault] has provided a prodigiously detailed set of possible descriptions whose main aim is . . . to overwhelm the individual subject or will and replace it instead with minutely responsive rules of discursive formation, rules that no one individual can either alter or circumvent. . . . Perhaps his in-

terest in rules is part of the reason why Foucault is unable to deal with, or provide an account of, historical change" (187-88).

26. Michel Foucault, "Maurice Blanchot: The Thought from Outside," in Maurice Blanchot and Michel Foucault, *Foucault/Blanchot* (New York: Zone Books, 1987), 55.

27. Ibid., 55, 57.

28. Ibid., 57.

29. Ibid., 58.

30. See William E. Connolly, *Politics and Ambiguity* (Madison: University of Wisconsin Press, 1987).

31. See Murray Edelman, *Constructing the Political Spectacle* (Chicago: University of Chicago Press, 1988).

32. Ibid., 1. Edelman discusses National Security as a key example throughout the book. See pp. 15ff.

33. Strobe Talbott, *Deadly Gambits: The Reagan Administration and the Stalemate in Nuclear Arms Control* (New York: Vintage, 1985), 273-74.

34. Irony is, suddenly, a popular topic in American social theory, perhaps in response to the prominent, neoliberal view espoused in Richard Rorty, *Consequences of Pragmatism* (Minneapolis: University of Minnesota Press, 1982) and *Contingency, Irony, and Solidarity* (Cambridge: Cambridge University Press, 1989). For responses to Rorty's quietist irony, see John Evan Seery, "Irony and Death (Or, Rorty contra Orpheus)" (unpublished paper), and his *Political Returns: Irony in Politics and Theory from Plato to the Antinuclear Movement.* See also Candace D. Lang, *Irony/Humor: Critical Paradigms* (Baltimore: The Johns Hopkins University Press, 1988).

35. Music and lyrics by Nick Lowe, "(What's So Funny 'Bout) Peace, Love, and Understanding," on Elvis Costello & The Attractions, "Armed Forces" (Columbia Records: New York, 1978). For an analysis of ironic sensibilities in popular culture, see George Lipsitz, *Time Passages: Collective Memory and American Popular Culture* (Minneapolis: University of Minnesota Press, 1990), especially his reading of Rahsaan Roland Kirk's combination of theatricality and the dark growl, "This ain't no sideshow" (3-21). Pop irony is examined in the context of popular music in Greil Marcus, *Mystery Train: Images of America in Rock 'n' Roll Music*, rev. ed. (New York: Dutton, 1982). Marcus sustains a major commentary on the Situationist roots of popular culture, with no little irony, in Greil Marcus, *Lipstick Traces: A Secret History of the Twentieth Century* (Cambridge, Mass.: Harvard University Press, 1989).

36. The congruence between Edelman's *Constructing the Political Spectacle* and more conventional "agenda" analyses of public policies is important because it imposes itself at precisely this point.

37. Seery, *Political Returns*, 294-99, 309-11, 320-25.

38. John Evan Seery, "Floating Balloons: An Essay on Nonviolent Theory, Irony, and the Anti-Nuclear Movement," *Soundings* 70 (Fall-Winter 1987): 373. See also Seery, *Political Returns*, 332-33.

39. Joseph Gusfield, *The Culture of Public Problems: Drinking-Driving and the Symbolic Order* (Chicago: University of Chicago Press, 1981), 173-96.

40. Ibid., 190.

41. Ibid., 193.

42. Rorty's ironic stance is analyzed in a different context in my "Our Babel: Dewey/Lippmann/Rorty," (unpublished paper).

43. Gusfield, *The Culture of Public Problems*, 176.

44. Jean Baudrillard, *The Evil Demon of Images* (Sydney, Australia: The Power Institute of Fine Arts, 1987), 15.

45. We might well date the collapse from Reagan's SDI proposal, a reaction to the freeze movement that instantaneously reversed and engulfed all opposition.

46. See Baudrillard, *The Evil Demon of Images*, 23.

47. Ibid., 51-52.

48. See R. Roy, "Limits of Genealogical Approach to International Politics," *Alternatives* 13 (January 1988): 77-83.

49. The interruption is "a place of the reinscription of the dialectic into deconstruction, . . . a necessary interruption which allows something to function." In such a "productive interruption, . . . we can pull together even if we bring each other to crisis." G. Spivak, "Practical Politics of the Open End: An Interview, Conducted by S. Harasym," *Canadian Journal of Political and Social Theory* 12 (no. 1-2, 1988): 51-69, 67-68.

50. J. W. Bernauer, "Michel Foucault's Ecstatic Thinking," in *The Final Foucault*, ed. by J. Bernauer and D. Rasmussen (Cambridge, Mass.: MIT Press, 1988), 45-82.

51. Michel Foucault, "The Ethic of Care for the Self as a Practice of Freedom: An Interview with Michel Foucault on January 20, 1984," conducted by R. Fornet-Betancourt, H. Becker, and A. Gomez-Müller, trans. J. D. Gauthier, in *The Final Foucault*, ed. Bernauer and Rasmussen, 3.

52. Michel Foucault, "The Ethic of Care for the Self," 5. See also Michel Foucault, *The Use of Pleasure: Volume 2 of The History of Sexuality* (New York: Pantheon, 1985), and *The Care of the Self: Volume 3 of The History of Sexuality* (New York: Pantheon, 1986).

53. Michel Foucault, "The Ethic of Care for the Self," 15.

54. Despite the unpopularity of pragmatism among poststructuralists, the model for this mode of political life is John Dewey, *Art as Experience* (New York: Minton, 1934). For a discussion of this aesthetics in the political context of environmentalism, see William Chaloupka, "John Dewey's Social Aesthethics as a Precedent for Environmental Thought," *Environmental Ethics* 9 (Fall 1987): 147-64.

55. Michel Foucault, "The Ethic of Care for the Self," 15.

## 6. Power/Cheekiness

1. Rux Martin, "Truth, Power, Self: An Interview with Michel Foucault, October 25, 1982," in *Technologies of the Self: A Seminar with Michel Foucault*, ed. Luther H. Martin, Huck Gutman, and Patrick H. Hutton (Amherst: University of Massachusetts Press, 1988), 15.

2. Diane Rubenstein, "The Anxiety of Affluence: Baudrillard and Sci-Fi Movies of the Reagan Era," in *Jean Baudrillard: The Disappearance of Art and Politics*, ed. William Stearns and William Chaloupka (New York: St. Martin's, 1991), 65.

3. Jean Baudrillard, *Cool Memories*, trans. Chris Turner (London: Verso, 1990), 23.

4. Murray Edelman, *The Symbolic Uses of Politics* (Urbana: University of Illinois Press, 1967), 114.

5. Calvin Thomas, "Baudrillard's Seduction of Foucault," in Stearns and Chaloupka, *Jean Baudrillard*, 137.

6. Peter Sloterdijk, *Critique of Cynical Reason* (Minneapolis: University of Minnesota Press, 1987), 101, 102.

7. Many of his critics have cited Baudrillard's propensity for exaggeration and wild statement. Robert Hughes cited Baudrillard's "remarkable silliness," decided that he had reduced politics to "mere bubbles on the surface of simulation," and published his critique next to a Levine caricature showing Baudrillard issuing bubbles from pursed lips, in Robert Hughes, "The Patron Saint of Neo-Pop," *New York Review of Books* 36 (June 1, 1989): 29-32. The publication of *America* met several such critiques, including Richard Vine, "The Ecstasy of Jean Baudrillard," *The New Criterion* (May 1989): 43.

8. Jean Baudrillard, *America*, trans. Chris Turner (London: Verso, 1988), 54.

9. Jean Baudrillard, "Transpolitics, Transsexuality, Transaesthetics," trans. Michel Valentin, in Stearns and Chaloupka, *Jean Baudrillard*, 26.

10. Baudrillard, *America*, 63. Emphasis added.

11. Comment by Baudrillard in Eric Johnson, ed., "Baudrillard Shrugs: A Seminar on Terrorism and the Media, with Sylvere Lotringer and Jean Baudrillard," in Stearns and Chaloupka, *Jean Baudrillard*, 302.

12. Arthur Kroker, "Panic Tocqueville," in his *Panic Encyclopedia: The Definitive Guide to the Postmodern Scene* (New York: St. Martin's Press), 86-91.

13. The Situationists probably express an aesthetic politics in terms closest to Baudrillard, who avers that he is "always a Situationist" in Johnson, "Baudrillard Shrugs." Another example of the aesthetic turn is found in Henry S. Kariel, *Beyond Liberalism* (New York: Harper Colophon, 1977).

14. Baudrillard, *America*, 126.

15. Edelman, *The Symbolic Uses of Politics*, 172.

16. This is what many political movements in recent decades have coded by the arrogant term "consciousness"; understanding the reversibility of representation is coded as a coming to consciousness. I would only argue that this discovery can be made less metaphysical and more political.

17. The term "fatal strategies" has not been without difficulty for Baudrillard and the U.S. publisher of most of his work, Sylvere Lotringer of Semiotext[e]. See Johnson, "Baudrillard Shrugs."

18. Baudrillard most clearly adopts this mood of melancholy in *Cool Memories*: "I have gone as far as I can in expressing sadness" (25).

19. Although Baudrillard and Peter Sloterdijk never mention each other, the two are linked in Andreas Huyssen's introduction to Sloterdijk's *Critique of Cynical Reason*.

20. Sloterdijk, *Critique of Cynical Reason*, 5.

21. Ibid., 111. His emphasis.

22. Ibid., 124.

23. Ibid., 130.

24. Johnson, "Baudrillard Shrugs.'

25. Timothy W. Luke, *Screens of Power: Ideology, Domination, and Resistance in Informational Society* (Urbana: University of Illinois Press, 1989), 90.

26. Ibid., 256.

27. Jean Baudrillard, *Seduction*, trans. Brian Singer (New York: St. Martin's Press, 1990).

28. See Martin, et al., *Technologies of the Self*, 3-16.

29. Baudrillard, "Transpolitics, Transsexuality, Transaesthetics," 15-16.

## 7. On Lastness: Nuclearism and Modernity

1. See Albert Camus, *The Rebel: An Essay on Man in Revolt* (New York: Vintage, 1956), 13-22.

2. Jacques Derrida, "No Apocalypse, Not Now (full speed ahead, seven missiles, seven missives)," *Diacritics* 14 (Summer 1984), 23.

3. See Spencer R. Weart, *Nuclear Fear: A History of Images* (Cambridge, Mass.: Harvard University Press, 1988).

4. Jean Baudrillard, "Forget Baudrillard: An Interview with Sylvere Lotringer," *Forget Foucault* (New York: Semiotext[e], 1987), 109-10.

5. There is evidence that some of this was understood by the freezers. See Waller's (50) report of his early discussion with defense analyst David Doerge, who explained that clearly "the most important thing about a freeze resolution would be its symbolic value . . . as a call

for action." Waller (30) also emphasizes that the freeze was designed to avoid the pitfalls of radical idealism. Douglas C. Waller, *Congress and the Nuclear Freeze* (Amherst: University of Massachusetts Press, 1987).

6. Donna Haraway, "A Manifesto for Cyborgs: Science, Technology, and Socialist Feminism in the 1980s," *Socialist Review* 15 (March-April 1985): 85-86.

7. Ibid., 87.

8. Ibid., 101.

9. Stanley Kubrick (writer and director), *Dr. Strangelove, or How I Learned to Stop Worrying and Love the Bomb* (Great Britain: Hawk Films, 1963).

10. R.E.M., "It's the End of the World As We Know It (And I Feel Fine)," *R.E.M. #5 Document* (Los Angeles: International Record Syndicate, 1987).

11. Avital Ronell, "Starting from Scratch: Mastermix," *Socialist Review* 18 (April-June 1988): 77.

# Selected Bibliography

## Books

Baudrillard, Jean. *For a Critique of the Political Economy of the Sign*. Translated by Charles Levin. St. Louis: Telos Press, 1981.

———. *Simulations*. Translated by Paul Foss, Paul Patton, and Philip Beitchman. New York: Semiotext[e], 1983.

———. *In the Shadow of the Silent Majorities . . . Or the End of the Social: And Other Essays*. Translated by Paul Foss, John Johnston, and Paul Patton. New York: Semiotext[e], 1983.

———. *The Evil Demon of Images*. Sydney, Australia: Power Institute of Fine Arts, 1987.

———. "Forget Baudrillard: An Interview with Sylvere Lotringer." In *Forget Foucault*. New York: Semiotext[e], 1987.

———. *America*. Translated by Chris Turner. New York: Verso, 1988.

———. *Cool Memories*. Translated by Chris Turner. New York: Verso, 1990.

———. *Fatal Strategies* . Translated by Philip Beitchman and W. G. J. Niesluchowski. Edited by Jim Fleming. New York: Semiotext[e], 1990.

———. *Seduction*. Translated by Brian Singer. New York: St. Martin's, Press, 1990.

Berman, Marshall. *All That Is Solid Melts into Air: The Experience of Modernity*. New York: Simon & Schuster, 1982.

Bernauer, James, and David Rasmussen, eds. *The Final Foucault*. Cambridge: MIT Press, 1988.

Connolly, William E. *Politics and Ambiguity*. Madison: University of Wisconsin Press, 1987.

Deleuze, Gilles, and Félix Guattari. *Anti-Oedipus: Capitalism and Schizophrenia*. Minneapolis: University of Minnesota Press, 1983.

———. *Nomadology: The War Machine*. New York: Semiotext[e], 1986.

Der Derian, James. *On Diplomacy: A Genealogy of Western Estrangement*. New York: Basil Blackwell, 1987.

Der Derian, James, and Michael J. Shapiro, eds. *International/Intertextual Relations: Post-modern Readings of World Politics*. Lexington, Mass.: Lexington Books, 1989.

Draper, Theodore. *Present History: On Nuclear War, Détente, and Other Controversies*. New York: Random House, 1983.

Dreyfus, Hubert. *What Computers Can't Do*. Rev. ed. New York: Harper & Row, 1979.

Dumm, Thomas. *Democracy and Punishment: Disciplinary Origins of the United States*. Madison: University of Wisconsin Press, 1987.

Edelman, Murray. *The Symbolic Uses of Politics*. Urbana: University of Illinois Press, 1967.

_____. *Constructing the Political Spectacle*. Chicago: University of Chicago Press, 1988.

Foucault, Michel. *The History of Sexuality: Volume 1: An Introduction*. New York: Vintage, 1978.

_____. *Discipline and Punish: The Birth of the Prison*. New York: Vintage, 1979.

_____. *Power/Knowledge: Selected Interviews and Other Writings, 1972-1977*. New York: Pantheon, 1980.

_____. *The Use of Pleasure: Volume 2 of The History of Sexuality*. New York: Pantheon, 1985.

_____. *The Care of the Self: Volume 3 of The History of Sexuality*. New York: Pantheon, 1986.

_____. *Politics, Philosophy, Culture: Interviews and Other Writings of Michel Foucault, 1977-1984*. Edited by Lawrence D. Kritzman. New York: Routledge, 1988.

Foucault, Michel, and Mauricek Blanchot. *Foucault/Blanchot*. New York: Zone Books, 1987.

Frankovits, André, ed. *Seduced and Abandoned: The Baudrillard Scene*. New York: Semiotext[e], 1984.

Gusfield, Joseph. *The Culture of Public Problems: Drinking-Driving and the Symbolic Order*. Chicago: University of Chicago Press, 1981.

Hersh, Seymour M. *"The Target Is Destroyed": What Really Happened to Flight 007 and What America Knew About It*. New York: Random House, 1986.

Kroker, Arthur. *Panic Encyclopedia: The Definitive Guide to the Postmodern Scene*. New York: St. Martin's Press, 1989.

Kroker, Arthur, and David Cook. *The Postmodern Scene: Excremental Culture and Hyper-Aesthetics*. New York: St. Martin's Press, 1986.

Lacan, Jacques. *The Structuralists from Marx to Levi-Strauss*. Garden City, N.Y.: Anchor, 1972.

Levy, Steven. *Hackers: Heroes of the Computer Revolution*. Garden City, N. Y.: Doubleday, 1984.

Lifton, Robert Jay. *Death in Life: Survivors of Hiroshima*. New York: Random House, 1967.

Lipsitz, George. *Time Passages: Collective Memory and American Popular Culture*. Minneapolis: University of Minnesota Press, 1990.

_____. *The Future of Immortality*. New York: Basic Books, 1987.

Loeb, Paul Rogat. *Nuclear Culture*. New York: Coward, McCann & Geoghegan, 1982.

_____. *Hope in Hard Times: America's Peace Movement and the Reagan Era*. Lexington, Mass.: D.C. Heath, 1987.

Luke, Timothy W. *Screens of Power: Ideology, Domination, and Resistance in Informational Society*. Urbana: University of Illinois Press, 1989.

Marcuse, Herbert. *One-Dimensional Man: Studies in the Ideology of Advanced Industrial Society*. Boston: Beacon Press, 1964.

Martin, Luther H., Huck Gutman, and Patrick H. Hutton, eds. *Technologies of the Self: A Seminar with Michel Foucault*. Amherst: University of Massachusetts Press, 1988.

O'Brien, Timothy. *The Nuclear Age*. New York: Knopf, 1985.

Papert, Seymour. *Mindstorms: Children, Computers, and Powerful Ideas*. New York: Basic Books, 1980.

Pirsig, Robert. *Zen and the Art of Motorcycle Maintenance: An Inquiry Into Value*. New York: William Morrow, 1974.

Rogin, Michael. *"Ronald Reagan," the Movie*. Berkeley: University of California Press, 1987.

Said, Edward. *The World, the Text, and the Critic*. Cambridge, Mass.: Harvard University Press, 1983.

Schell, Jonathan. *The Fate of the Earth*. New York: Knopf, 1982.

_____. *The Abolition*. New York: Knopf, 1984.

Shapiro, Michael. *Language and Political Understanding*. New Haven: Yale University Press, 1981.

_____. *The Politics of Representation: Writing Practices in Biography, Photography, and Policy Analyses*. Madison: University of Wisconsin Press, 1988.

Shapiro, Michael, ed. *Language and Politics*. New York: New York University Press, 1984.

Sloterdijk, Peter. *Critique of Cynical Reason*. Minneapolis: University of Minnesota Press, 1987.

Solomon, J. Fisher. *Discourse and Reference in the Nuclear Age*. Norman: University of Oklahoma Press, 1988.

Stearns, William, and William Chaloupka, eds. *Jean Baudrillard: The Disappearance of Art and Politics*. New York: St. Martin's Press, 1991.

Talbott, Strobe. *Deadly Gambits: The Reagan Administration and the Stalemate in Nuclear Arms Control*. New York: Knopf, 1984.

Turkle, Sherry. *The Second Self: Computers and the Human Spirit*. New York: Simon & Schuster, 1984.

Virilio, Paul. *Speed and Politics: An Essay on Dromology*. New York: Semiotext[e], 1986.

_____. *War and Cinema: The Logistics of Perception*. London: Verso, 1989.

Waller, Douglas C. *Congress and the Nuclear Freeze*. Amherst: University of Massachusetts Press, 1987.

Weart, Spencer R. *Nuclear Fear: A History of Images*. Cambridge, Mass.: Harvard University Press, 1988.

Wills, Garry. *Reagan's America: Innocents at Home*. Garden City, N.Y.: Doubleday, 1987.

Winner, Langdon. *Autonomous Technology: Technics-out-of-Control as a Theme in Political Thought*. Cambridge: MIT Press, 1977.

## Articles

Cooper, Barry. "The Meaning of Technology at the End of History." *University of Minnesota Center for Humanistic Studies Occasional Papers* 7 (1986).

Derrida, Jacques. "No Apocalypse, Not Now (full speed ahead, seven missiles, seven missives)." *Diacritics* 14 (Summer 1984): 20-31.

Ferguson, Frances. "The Nuclear Sublime." *Diacritics* 14 (Summer 1984): 4-11.

Foss, Paul. "Despero Ergo Sum." *Seduced and Abandoned: The Baudrillard Scene*, edited by André Frankovits, 9–16. New York: Semiotext[e], 1984.

Haraway, Donna. "A Manifesto for Cyborgs: Science, Technology, and Socialist Feminism in the 1980s." *Socialist Review* 15 (March-April 1985): 64-107.

Jameson, Fredric. "Postmodernism or the Cultural Logic of Late Capitalism." *New Left Review* 146 (July-August 1984): 66-67.

Kariel, Henry S. "Affirming a Politics of Inconsequence." *Polity* 17 (Fall 1984): 145-60.

Kateb, George. "Thinking about Human Extinction (I): Nietzsche and Heidegger." *Raritan* 6 (Fall 1986): 1-28.

_____. "Thinking about Human Extinction (II): Emerson and Whitman." *Raritan* 6 (Winter 1987): 1-22.

Klein, Richard, and William B. Warner. "Nuclear Coincidence and the Korean Airline Disaster." *Diacritics* 16 (Spring 1986): 2-21.

McCanles, Michael. "Machiavelli and the Paradoxes of Deterrence." *Diacritics* 14 (Summer 1984): 12-9.

MacCannell, Dean. "Baltimore in the Morning . . . After: On the Forms of Post-Nuclear Leadership." *Diacritics* 14 (Summer 1984): 34-46.

Ronell, Avital. "Starting from Scratch: Mastermix." *Socialist Review* 18 (April-June 1988): 74-85.

Rubenstein, Diane. "Hate Boat: Greenpeace, National Identity, and Nuclear Criticism." In *International/Intertextual Relations: Postmodern Readings of World Politics*, edited by James Der Derian and Michael J.Shapiro, 231–56. Lexington, Mass.: Lexington Books, 1989.

_____. "Food for Thought: Metonymy in the Late Foucault." In *The Final Foucault*, edited by James Bernauer and David Rasmussen, 83-101. Cambridge: MIT Press, 1988.

_____. "The Mirror of Reproduction: Baudrillard and Reagan's America." *Political Theory* 17 (November 1989): 582-606.

Schwenger, Peter. "Writing the Unthinkable." *Critical Inquiry* 13 (Autumn 1986): 33-48.

Spivak, G. "Practical Politics of the Open End: An Interview, Conducted by S. Harasym." *Canadian Journal of Political and Social Theory* 12 (no. 1-2, 1988): 51-69.

Walker, R. B. J. "Genealogy, Geopolitics and Political Community: Richard K. Ashley and the Critical Social Theory of International Politics." *Alternatives* 13 (January 1988): 86.

_____. "*The Prince* and 'The Pauper': Tradition, Modernity, and Practice in the Theory of International Relations." In *International/Intertextual Relations: Postmodern Readings of World Politics*, edited by James Der Derian and Michael J. Shapiro, 25–48. Lexington, Mass.: Lexington Books, 1989.

Walsh, Birrell. "The Monkey Trap or the Mystical Engine?" *Whole Earth Review* 44 (January 1985): 1-2.

Willis, Ellen. *Village Voice*, April 30, 1979: 8.

Wills, Garry. "Critical Inquiry (*Kritik*) in Clausewitz." *Critical Inquiry* 9 (December 1982): 287.

_____. "Aerie Visions." *New York Review of Books*, (March 17, 1988): 7.

# Index

absence, 26, 28-32, 38-42, 133
accident, 13-16
agency, 13, 44, 66
alienation, 91
ambiguity, 128
"Amerika," 9
apathy, 91
*aporia,* 30, 72, 95, 140 n. 23
Aristotle, 11
arms control, 38, 69
arms race, 72
Artificial Intelligence, 43, 54, 59
Asimov, Isaac, 58
Asimov Priorities, 145 n. 22
*Atomic Cafe,* 17, 18
attitude, 115
author, xv, 95, 96, 113, 115
automation, 58
autonomy, 44, 57

Babbage, Charles, 57
barracks, 25, 27
battle 26, 28-32
Baudrillard, Jean, 2, 17-20, 33-35, 40-41,
    48, 57-58, 60, 65, 73, 84, 90-92, 101,
    106-16, 121-24, 135, 152 n. 7
Berlin Wall, 23
Berman, Marshall, 84

Bernauer, James W., 102
black box, 52
Bloom, Allan, 91
body, 27, 28, 45, 56, 62, 129, 132
bomb, 33, 44-49, 61, 63, 66, 67, 117, 131,
    137
border, 45
boundary, 128
boxing, 128
Bush, George, 123, 124, 131
Byron, Lord, 55

C3I (command-control-communication in-
    telligence), 44
calculation, 49, 58
Caldicott, Helen, 114
Camus, Albert, 127, 150 n. 12
Cheyenne Mountain, 53
*China Syndrome,* 79
citizen, 26, 113, 115, 122
citizenship, 98
clarity, 44, 53
Clausewitz, Karl von, 15, 31, 32
code, 2, 45, 68, 94, 97, 98, 107, 124
Cohen, Leonard, 114
Cold War, 23, 24, 35, 87
commodity fetishism, 88
communication, 119

computer, 43, 62, 65-67, 137
control, 16, 55-58, 62, 66, 74, 124, 143
    n. 31
Cook, David, 65
Cooper, Barry, 7
Costello, Elvis, 98
counterforce, 74
cowboy, 147 n. 12
Cronkite, Walter, 130
cyborg, 44, 81, 136-38
cynicism, 116, 117, 123
Czechoslovakia, 124

danger, 15
death, 8, 63
deconstruction, xi, xiii, xvi, 5, 8, 10, 30, 67,
    91, 97, 125-28
Deleuze, Gilles, 26, 61-65
democracy, 5
demonstration, 125
Der Derian, James, 38, 39
Derrida, Jacques, xiv, 2, 4, 8, 9, 10, 11, 17,
    30, 72, 95, 101, 109, 119, 131, 132,
    140 n. 21
Desert Bloom, 64, 146 n. 39
desire, 60-64, 77
deterrence, 12, 13, 17, 21, 27, 29-35, 60,
    68-71, 74, 76, 82, 83, 133
Dewey, John, 112, 127, 144 n. 42, 152
    n. 54
Diacritics, 2, 140 n. 20
Diogenes, 110, 116, 117
diplomacy, 39
disappearance, 28, 30
discipline, 16, 18, 25-27, 36, 37, 56, 58,
    135
Dreyfus, Hubert, 55, 59
Dr. Strangelove, or How I Learned to Stop
    Worrying and Love the Bomb, 137, 146
    n. 35

economics, 57
Edelman, Murray, 70, 108, 115
Einstein, Albert, 16, 20, 130
electronic toys, 53
elites, 87
Ellul, Jacques, 56
Enlightenment, 57, 117
entrepreneurial attitude, 48
euphemism, 19-22
EVERLAST, 128-29

factory, 57, 136
fallout, 34
FBI, 38
Felsenstein, Lee, 50
feminism, 40, 96, 136
Ferguson, Frances, 10
flag burning, 46
foreign policy, 86-88, 94, 95
Forsberg, Randall, 35
Foss, Paul, 41
Foucault, Michel, 2, 4, 7, 11, 17, 21, 25, 32,
    35-37, 66, 76, 90, 102, 103, 108, 109,
    122, 123, 128; on control of bodies 56;
    Discipline and Punish, 25-26; on ecol-
    ogy movements, 102; on ethics, 102; on
    humanism, 106; on language, 94-96; on
    the local, 86; and the modern self, 37;
    on the specific intellectual, 5-6; on ther-
    apy, 40
fractal, 123, 124
Frankenstein, 56, 58, 145 n. 20
freedom, 106

games, 12
genealogy, 32
Germany, 82
God, 3, 5, 41
Gorbachev, Mikhail, 12, 35, 40, 47, 49, 69,
    75, 83-85, 107, 111, 125, 144 n. 6
Greenpeace, 60
Greens, 89
Guattari, Félix, 26, 61-65
guerrilla warfare, 26, 28, 31, 32
gun control, 74
Gusfield, Joseph, 99-100

hackers, 51-52, 55
Hall, Fawn, 144 n. 36
Haraway, Donna, 44, 53-54, 136
Hartz, Louis, 114
Heidegger, Martin, 4
Heisenberg, Werner, 11
Heron, Gil Scott, 130
Hersh, Seymour, 29, 60
Hiroshima, 8, 16, 18, 46, 68, 130
Hoover, J. Edgar, 123
human nature, 36
humanism, xiv, xv, 5, 21, 24, 36-37, 40, 62,
    65, 66, 105-8, 128
Huyssen, Andreas, 43, 66

IBM, 47, 50
idealism, 5
ideology, 1, 121-22
image, 8
individualism, 4-5
intellectuals, 5, 6, 91, 121, 127, 131
intertextuality, 78
intervention, 41, 42, 86, 93, 94, 101, 102, 138
Iran Air flight, 655, 13
Iraq, 124
irony, 44, 45, 97-103, 109, 113, 115, 151 n. 34; sociological, 99

Jameson, Fredric, 79

Kateb, George, 3-6, 75, 106, 140 n. 12
Klein, Richard, 14-16, 141 n. 39
knights, 26
Korean Air Lines (KAL) flight 007, 14, 29, 141 n. 39
Kroker, Arthur, 65, 66, 112
Kubrick, Stanley, 63, 137
Kuwait, 82
kynics, 116-19, 125

Lacan, Jacques, 60, 63
Lang, Fritz, 66
language, 87, 94-97, 101, 108, 121, 128, 138, 148 n. 23
leukemia, 17, 33, 144 n. 42
liberal, xiv, 25, 65, 89
lifestyle, 88-89, 92-96, 99, 103
Lifton, Robert Jay, 3, 8, 114; on "numbing," 17
Lipsitz, George, 147 n. 11, 151 n. 35
literary criticism, xiv, xvi, 11, 12, 19, 44
literary production, 83
Locke, John, 98
LOGO, 51, 62
Los Alamos, N. M., 46
Lotringer, Sylvere, 118
Luke, Timothy, 31, 121-22
Lyotard, Jean-François, 118

McCanles, Michael, 12, 30, 74
MacCannell, Dean, 63, 146 n. 36
Machiavelli, Niccolò, xiv, 2, 12, 84, 148 n. 23
Marcus, Greil, 151 n. 35
Marcuse, Herbert, 20

Marines, and TV ads, 30
Marx, Karl, xiv, 7, 25, 39, 57, 58, 65, 84, 89, 91, 107
masses, 90, 91
Maytag repairman, 32, 143 n. 22
Merleau-Ponty, Maurice, 127
metaphor, 60
metaphysics, xiv, 3, 27, 33, 67
metonymy, 60-62, 111, 115, 130
Metropolis, 66
militarism, 34, 46
military obligation, economy of, 31
miniaturization, 54
Minsky, Marvin, 52
mirror, 63, 66, 83
missiles, 26-29, 60, 97
models, 12, 93, 100, 106
modernity, xiii, 68, 84, 117, 134
modesty, 98
monkey trap, 67

National War College, 29
nature, 6, 17, 45
negation, 7
New Age, 102
Nietzsche, Friedrich, 3, 58, 65, 91
nihilism, 41, 58, 134
1989, 23, 85, 105, 107, 111, 118, 124
Nixon, Richard, 90, 112, 122
North, Oliver, 38, 143 n. 31
nuclear criticism, xiv, xvi, 2-5, 8, 10, 17-20, 74
nuclear fallout, 34
nuclear freeze movement, 35, 69, 70-75, 81, 82-87, 136, 153 n. 5
nuclear terminology, xv
nuclear therapy, 17
nuclear war, 12, 27
nuclear winter, 71
nuclearism, xiii, xv, 1, 2, 17, 33, 37, 44, 45, 99, 125, 126, 133
"nuke" (defined), xvi
nukespeak, 19-21

opacity, 44-47, 52, 59-62, 67
open secret, 114
Orr, Verne (Air Force Secretary), 30

panopticon, 39, 122
Papert, Seymour, 51
parades, 32, 46

paradox, 16, 17, 30, 58, 68, 71, 74, 76, 109, 126
Pentagon, 30, 60, 80
Pershing II, 60
Pfautz, Major General James C., 29
Pickens, Slim, 137
Pirsig, Robert, 146 n. 27
Plato, 130
political culture, 50
political science, 107, 108, 114, 133, 140 n. 21
politics, 2, 45, 56, 70-72, 80, 86, 89, 93, 95, 98, 101, 108, 113, 115, 122, 125, 131, 134
popular art, 98
popular culture, 43
popular will, 105
populism, 50, 52
postmodernism, xiii, xiv, 41, 42, 134, 135
power, xii, xiv, 10, 17, 25, 37, 40, 56, 68, 87, 94, 102, 104-9, 115-19, 132-35

R.E.M., 138
Rambo, 24, 32
Rather, Dan, 131
rationalism, 109
Reagan, Ronald, 12, 69-71, 74-77, 80-84, 97, 107, 112, 123, 125, 133, 144 n. 6; administration, 14, 69; era, 24; fundamentalism and, 70; popularity, 70
real, the, 118, 132, 134
recycling, 93
referent, xiv, 87
representation, 62, 69, 70, 91-94, 111, 115, 125, 148 n. 22
responsibility, 14
reversibility, 13, 94, 107, 111, 122
revolution, 27, 35, 79, 134, 135
rhetoric, 8, 10, 21, 25, 31, 63, 79, 109, 112
Richland (Wash.) High School, xiii, xvi, 46
Robocop, 58, 60
robots, 43-45, 48, 50, 54-66, 77, 80-82, 137
Rogin, Michael, 69, 81
Ronell, Avital, 28
Rorty, Richard, 100, 151 n. 34
Rubenstein, Diane, 60, 69, 107
rules, 94

Said, Edward, 19
Schell, Jonathan, 3, 46

school, 51, 64
Schwenger, Peter, 8
science fiction, 43, 58
Scorsese, Martin, 129
SDI, 15, 47, 73-84; "Astrodome" ad for, 81
Searle, John, 55
Seery, John Evan, 99, 150 n. 11, 151 n. 34
Selective Service System, 31
self, 37, 57, 60, 88, 89, 95-97, 134
Shapiro, Michael, 148 n. 22
Shelley, Mary Wollstonecraft, 56
silent majority, 90, 91, 112
Silliman, Ron, 11
simplicity, 93
simulation, 28, 65, 90, 107, 143 n. 25
Sloterdijk, Peter, 27, 28, 110, 116-19, 122, 123
Smith, Adam, 90
social, the, 90, 93, 96, 97
social science, 92, 133
sociology, 82
soldiers, 25, 28
Solomon, J. Fisher, 10-11, 140 n. 20
"solution" talk, 135
species, 6-9
spectacle, 70, 96
speed, 26-27, 44, 68, 72, 76, 82, 107, 120
Spivak, Gayatri Chakravorty, 102, 152 n. 49
spy, 15, 38, 39, 122-23, 141 n. 39
spy satellites, 38
stability, 12, 13
"Star Trek," 80
Star Wars, 44, 77-81
stereotype, 77, 82
Strategic Air Command, 47, 53
students, 91, 92
surveillance, 26, 36-40, 56, 122, 133, 135
survival, xiv, 1, 2-7, 12, 19-22, 37, 38
symbol, xii, 1, 2, 98

taboo, 9, 46, 66
tanks, 27
teletradition, 31
television, 30, 41
Teller, Edward, 20
territory, 27, 29, 82, 120
texts and textuality, 9, 10, 50, 74, 140 n. 21, 148 n. 22, 149 n. 4
"The Day After," 9, 133
theology, 3

therapy, 40-43, 68, 89, 142 n. 49
Thomas, Calvin, 109
Thompson, Hunter S., 119
Three Mile Island, 13, 48-49, 79
Tiananmen Square, 23, 124, 125
totality, 3, 4, 130
Transformer toy, 54
truth, 95, 102, 103, 106, 110, 130, 132
Turkle, Sherry, 53
*2001: A Space Odyssey,* 80

unspeakability, 7-11, 18-21, 73
utopia, 111

values, xiii, xiv, 5
verification, 38
video games, 49
Vietnam veterans, 32
Vietnam War, 34, 130
Virilio, Paul, 26, 28, 72, 76, 82; *War and Cinema,* 28

Walker, R. B. J., 87, 148 n. 23
Walsh, Birrell, 67
war, 26, 33, 61
war games, 30
war toys, 28
Warner, William B., 14-16, 141 n. 39
warrior, the, 24-33, 35-46, 117
Wells, H. G., 8
white trains, 47, 144 n. 7
Willis, Ellen, 48, 144
Wills, Garry, 15, 31, 76, 81, 147 n. 12
window of vulnerability, 15
Winner, Langdon, 44
Wittkopf, E. R., 87
*Wizard of Oz,* 53
work, 58, 136
World War II, 12

William Chaloupka is associate professor of political science at the University of Montana. He is the coeditor, with William Stearns, of *Jean Baudrillard: The Disappearance of Art and Politics*, and has published several articles on ethics and theoretical problems in the practice of politics.